Also by Dr Hilary Jones

Before You Call the Doctor
What's Up Doc?

A Day in Your Life

Dr Hilary Jones

CORGI BOOKS

TRANSWORLD PUBLISHERS
61–63 Uxbridge Road, London W5 5SA
A Random House Group Company
www.transworldbooks.co.uk

A DAY IN YOUR LIFE
A CORGI BOOK: 9780552159517

First published in Great Britain
in 2013 by Bantam Press
an imprint of Transworld Publishers
Corgi edition published 2014

The information in this book is believed to be correct as at May 2013 but is not to
be relied on in law and is subject to change. The author and publishers disclaim,
as far as the law allows, any liability arising directly or indirectly from the use, or
misuse, of any information contained in this book.

A CIP catalogue record for this book
is available from the British Library.

Addresses for Random House Group Ltd companies outside the UK
can be found at: www.randomhouse.co.uk
The Random House Group Ltd Reg. No. 954009

The Random House Group Limited supports the Forest Stewardship Council®
(FSC®), the leading international forest-certification organisation. Our books
carrying the FSC label are printed on FSC®-certified paper. FSC is the only
forest-certification scheme supported by the leading environmental organisations,
including Greenpeace. Our paper procurement policy can be found at
www.randomhouse.co.uk/environment

Typeset in 11/15pt Minion by Falcon Oast Graphic Art Ltd.
Printed and bound by CPI Group (UK) Ltd, Croydon, CR0 4YY.

2 4 6 8 10 9 7 5 3 1

MIX
Paper from
responsible sources
FSC® C016897

For Annabella. Don't scare me like that ever again.

Introduction

THIS BOOK IS all about you, or to be more exact, about your body and how it ticks.

It will tell you all about the utterly incredible things your body does all on its own, without you ever being conscious of it. It will explain why you dream, and what exactly it is that makes you wake up in the morning and fall asleep at night. It tells you why you yawn or laugh, and how you see, hear and taste. It describes how your breathing and digestion simply happen without you even having to think about it and how your heart, no bigger than the size of your clenched fist, unfailingly goes on beating billions of times during your lifetime without ever getting tired.

There are so many physiological curiosities. What, for example, is a memory? Could a thought, a small parcel of chemicals travelling between nerve cells in your brain, theoretically be captured and injected into somebody else?

Are out-of-body experiences just tricks of our altered

consciousness? Can normal people really be turned into killers as a result of strokes or head injuries?

How many times a day does the average person break wind, what is snot and why can't some people empty their full-to-bursting bladder just because someone else is within earshot?

How do your bones repair themselves when they break and how does your blood clot when you are injured but not when it circulates around our body?

At what stage does an unborn baby begin to move and when does it become aware of the world around it? How do its brain, heart and limbs grow from just two basic cells and is its personality and sexual orientation already encoded at birth?

What exactly is love, and sexual desire, and how do people who are born deaf silently voice their thoughts inside their heads if it isn't in the form of words that they can never have heard?

Why do we become overweight, depressed or hungover? What processes make us grow old or develop cancer, and what goes on within our bodies as we approach the moment of death?

A Day in Your Life describes a typical day in the life of the Enniman family – a father, a pregnant mother, their two teenage children, their five-year-old daughter and the family dog – and it focuses on everything that goes on inside their bodies during that 24-hour period.

But everything that happens to them also happens to you. I've been a doctor in General Practice and on TV for over

35 years, learning new things every single day. But I've never learned as much as I have whilst writing this book and I've had enormous fun in the process. I hope it opens your eyes to the wonders of the human body and puts the 'fizz' back into physiology.

Chapter 1

THE HUMAN EYE is a remarkable organ and one of the most complex in our body, second only to the brain itself. At just 24 millimetres or one inch in diameter, and weighing about 28 grams, it boasts around 2 million separate working parts and can process 36,000 bits of data every hour. Our eyes can function at 100 per cent ability at any time of the day or night without rest, and are the only organs which can do so. Even when our eyes are gritty and we feel sleepy it is only the misfiring of the tired brain that affects our vision, as it fails to interpret the faithful signals from our eyes. We can determine speed and distance accurately enough to catch a fast-moving ball. We can read small print close up and thread the narrowest of cotton fibres through the tiniest eye of a needle. We can see a candle at a distance of 14 miles, and detect a vast range of colours, including 500 different shades of grey.

But right now, as Adam lies dozing in his bed, his eyes are shut, and the only sign of life is the darting side-to-side movement of the front of his eyes behind his closed lids. This is because he is in rapid-eye-movement sleep (REM), where his brain activity is increased and information gathered over the previous few days is processed to assimilate memories and experiences.

There are two main types of sleep which alternate through the varying cycles of a typical night's rest. Adam had his deepest sleeps during the fourth stage of non-REM sleep at roughly one hour and three hours after dropping off the previous night. He had a brief period of REM sleep in between, where he enjoyed some wonderful dreams, followed by yet more episodes of REM dreaming at increasingly frequent intervals throughout the night, each one lasting longer. Right now, at 6.30am, he is experiencing his last bout of REM, and unfortunately his dream is rather alarming. For some reason, he's gone right back to his schooldays and is re-enacting Sophocles' play *Oedipus Rex*. In this Ancient Greek tragedy, the Theban king, Oedipus, has unwittingly married his mother and killed his father. Here he is now, popping up in Adam's dream, trying to cleanse himself of his mortal sin by gouging out his eyes with his mother's golden brooches. Bloodstained fluid is pouring down both cheeks and he is screaming.

At that moment Adam wakes up and realizes the alarm on his mobile phone is doing exactly what he set it to do just before diving under the duvet last night. Funny, he is thinking, how the piercing ring of his alarm was incorporated into the last bit of his dream, even though his dream appeared to have been going on for some considerable time. But the dream cannot anticipate the alarm call; it can only incorporate the auditory signal within a few seconds of its being heard. How bizarre that his brain can perceive a dream to have evolved slowly when actually it is just a momentary reaction to extraneous sound. How strange that in his sleep he can so easily misinterpret the passage of time and create a whole chain of imaginary events.

Many scientists believe that it is this very dissociation between time and extraneous sensory input during sleep that explains the bizarre phenomenon of out-of-body experiences and consciousness under anaesthesia. As we awaken, a dream can seem protracted, vivid and utterly real. In reality it is purely a confusion of sensory inputs into the drowsy and semi-comatose brain.

As Adam gropes across to switch off the alarm, his mind wanders back to poor Oedipus and his ruined sight. Eyes are such sensitive and essential organs. Vision overshadows all our other senses and dominates our perception of consciousness from the moment we are born. Our eyes are sophisticated biological video cameras that focus light from our environment on to the sensitive film at the back of our eyes, producing chemical reactions that convert to electrical signals, which are then transmitted to the brain for highly complex processing. The result: three-dimensional vision. So why would anyone, let alone King Oedipus, want to pluck out their own eyes? Yet it happens.

Adam can dimly recollect a story in the newspaper about self-enucleation of the eyes being an extreme but fortunately rare form of self-harm. The article talked about a 48-year-old man who was brought to A&E with a history of self-gouging of both eyes. Both his orbits were bandaged and there was fluid oozing down his cheeks. He was calm and not in any apparent pain. His relatives reported that he had a troubled family life but no significant ocular history. He was known to be epileptic and apparently had had a seizure just before the self-enucleation. His medication included two anticonvulsants. What happened to him? Oh yes, now Adam remembers. He was admitted to the ward for observation. The poor man refused any closer examination of his eyes. A psychiatrist who saw him concluded that he had been suffering from post-ictal psychosis – a temporary form of insanity resulting from his epileptic seizure. The article went on to say that the eye sockets had been left to heal spontaneously with just antibiotic ointment to prevent infection. He was subsequently registered fully blind and later referred to an artificial eye centre for a prosthesis. Not surprisingly, it was expected that he would remain under psychiatric care.

These are the kind of awful thoughts people sometimes have in that twilight time between sleep and alertness. There is a theory that sometimes turning over on to your other side will make the intrusive ideas go away and that going back to sleep on the same side will make the same dream return. Sometimes the best thing is just to wake up and get on with the day.

Now Adam comes to think of it, his eyes do feel a bit gritty, and the lids are a little sore and dry. Is that why he had the dream? Did burning lids inform his brain and his brain in turn incorporate that into the dream? Is it a primitive reflex, perhaps alerting him to the possibility that something or someone might be trying to poke out his eyes? Who knows? All he is sure of is that he has to get up.

His eyes are still shut, but that isn't stopping the daylight filtering through the delicate skin of his eyelids, making him aware of the pinkish glow of the world beyond. His eyes are actually 'seeing' the pinkness of his blood as it flows through the tiny capillaries there. Slowly, his lids flutter open, and he takes his first furtive look around the room. Familiar room. Usual curtains. Mess on the floor. Regular sleeping partner. No burglar standing over him ready to plunge in a knife. Phew, what a relief. His brain has made its first personal security check of the day. He is fully awake.

What exactly is it that made Adam become conscious? How is this intangible phenomenon linked to a palpable physical organ like his brain? Weighing in at about 1.5 kilos, or 3.25lb, his brain looks and feels like hardened jelly. It's about the size of a cauliflower, yet the size rarely correlates with IQ and every human brain, whatever its volume and weight, contains more or less the same number of nerve cells and connections.

Consciousness is like nothing else we know. Thoughts, emotions and concepts are fundamentally different from anything else that makes up our environment and, for that matter, the universe. We can't touch, see or measure the consciousness within our mind. We can't place it anywhere in space or time.

Different levels of consciousness can direct our attention

towards the physical outside world or the inner world of our ideas and thoughts. We can focus on many disparate objects or perceptions at the same time, or home in on just one in fantastic detail. We can also experience varying depths of awareness, depending on whether awareness is fleeting and transient, or registered and encoded in memory for recall in the future. So whereabouts in the brain does that conscious awareness come from? Is any one bit more important than another?

In terms of the awakening of the brain and the arousal of conscious thought, nerve cell activity in the outer part of the cerebrum, especially in the frontal lobes of the brain, is probably most significant. When a signal from the rest of your body first registers in the deeper parts of your brain, it takes about half of one second for you to become conscious of that signal, when it is relayed to the 'higher' areas in the cerebral cortex. This is the outer layer of the brain, characterized by convolutions and fissures piled up tightly against one another as if this part of the brain was continually trying to expand but had nowhere to go inside the rigid box of the bony skull. Behind the forehead sits the frontal cortex, which is only active when any kind of experience becomes a conscious one. It is the cerebral cortex – the whole outer layer – that distinguishes humans from other animals and gives us the invaluable gift of consciousness and intelligence. Having evolved over millions of years, we have a disproportionately large cerebral cortex compared to other animals, with only dolphins and elephants coming anywhere near.

In the deeper part of your brain, the brain stem itself, lies an area known as the reticular activating system or RAS. It is this area that receives incoming sensory messages about the outside environment from the rest of your body and relays it onwards to the cortex to rouse it and prime it for action. This is your primary arousal mechanism. This is what wakes you up in the morning in response to your alarm call, a full bladder, or your daughter screaming after having a nightmare. It also helps to regulate the autonomic nervous system, a part of your nervous system over which you have no voluntary control, and it plays an important role too in controlling how fast your heart rate is and in regulating breathing. It influences many other essential bodily functions too, such as digestion, sweating, peeing and even sexual arousal. The signals coming right now from Adam's full bladder go a long way to explaining why he quite often wakes up with a morning erection. It is mostly down to his RAS.

But how did his eyes open by themselves like that? And how did they begin to discern the shapes and features of his bedroom, which is in almost total darkness?

The principal voluntary muscle of the upper eyelid is the levator palpebrae superioris. This is the muscle that needs to contract in order for you to open your eyes and gaze at the world in front of you. But there are small fibres of the orbicularis oculi and involuntary tarsal muscles, which are also found in the network of muscles in the eyelids. So when you frown, screw up your eyes or squint at light, your eyelids themselves will be included in the action. When you want to open your eyes for the first time in the morning, your brain sends a signal down the third cranial nerve, the oculomotor nerve, which runs a tortuous anatomical route from its origin deep within the midbrain down into the bony eye socket.

You thought opening your eyes would be the simplest thing in the world, but listen up, this is only the beginning. This nerve on its own opens the eyes and allows them to look upwards, inwards towards your nose, downwards and to the left and right. Two completely separate nerves are required to enable our eyes to look in other directions. In addition, they all have to operate in harmony. This whole process is incredibly intricate and complicated, but just another function of the human body that we completely take for granted.

Sometimes we only realize just how much we rely on our body's automatic functions when their normal function is interrupted by injury or ill health. When parts of our body don't work efficiently, we certainly know about it. A crippling bad back. A broken leg. A frozen shoulder. An abnormal rhythm of the heart. What if you couldn't properly open your eyes in the morning? This is something you do every single day of your life. You need to be able to do it to see the world. But for some people it's a real problem.

Drooping of the upper eyelid due to weakness in the levator palpebrae superioris is a condition known as ptosis. This can be the result of a disorder in the muscle or the nerve that controls the eyelid. A sagging lid can completely close or just partially close the eye and can be found on one or both sides. Very occasionally, it's present from birth and unless it's corrected it can prevent vision from developing normally.

When it starts for the first time in adults, it can occur as a result of the ageing process; as a symptom of a rare illness known as myasthenia gravis where antibodies attack and slowly destroy the receptors in the muscles that receive nerve impulses; or very occasionally, as a result of a brain tumour or a balloon-like deformity of a blood vessel in the brain known as an aneurysm. Increasingly these days, you might read about it as an unwanted side effect of a Botox injection, used cosmetically to get rid of wrinkles and frown lines on the forehead. If some of the botulinum toxin in the injection goes astray, instead of paralysing the muscles in the forehead that make you frown and give your face expression, partial paralysis of the muscle responsible for lifting the eyelid may occur, causing it to sag.

But none of this applies to Adam right now. He is awake. Both his eyes are open, and it's time to face the day. His pupils have been gradually dilating to take in as much of the dim light available in the bedroom as possible. But when he glances at the bright shaft of daylight pouring through the gap in the curtains, his pupils automatically constrict a little, and even more dramatically when he fumbles for the bedside light and finally switches it on.

The pupils are highly sensitive to light, opening up in the dark, or when you're excited, to allow more light into the eyes, but rapidly constricting to protect them from dazzling light, or when you are relaxed. The size of the pupil is controlled by the muscles of the iris, the coloured part of the eye surrounding it. Within the iris, there are circular muscles immediately adjacent to the pupil that run in a ring around it. There are also radial muscles which run at right angles to the circular muscle from the inner to the outer border of the iris, like the rays of the sun or the spokes of a bicycle wheel. In low light conditions, the circular muscles around the pupil relax, while the radial muscles contract to widen the pupil and allow more light to enter. Conversely, in bright light, the radial muscles on the outside of the iris relax, and the circular muscles next to the pupil contract, narrowing the pupil and reducing the amount of light coming into the eye.

But the eyes are even more sophisticated than this, because they work not only independently, but con-centrically as well. If a torch is shone into one eye, not only will the pupil of that eye contract, but the pupil of the other

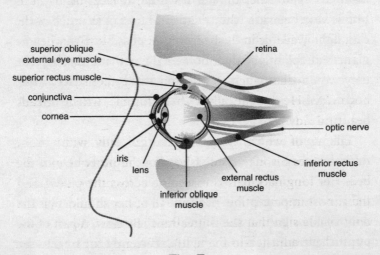

superior oblique
external eye muscle

superior rectus muscle

conjunctiva

cornea

iris

lens

inferior oblique
muscle

retina

optic nerve

inferior rectus
muscle

external rectus
muscle

The Eye

eye will also contract. It even happens if you're blindfolded over one eye. It is a protective reflex, and is dependent on a complicated cascade of nerve cell reactions occurring at different levels of the brain and nervous system. Overseeing the whole operation is the activity of the involuntary nervous system, which will influence the size of the pupils by means of tiny pockets of chemicals released at nerve endings, which cause the muscles to contract or relax. The type of chemicals released changes according to whether we're fearful or calm, anxious or relaxed. When we are afraid, hormones like adrenaline kick in and our pupils widen and dilate. When we're relaxed, our pupils are more likely to be constricted.

This is why recreational drugs like amphetamine, which mimic the effect of adrenaline, will dilate the pupil and why morphine and other opiates will have the opposite effect. Many years ago, in the court of Louis XIV of France, belladonna eye drops, originally derived from the deadly nightshade plant, were fashionable with the ladies because they would dilate their pupils and make them look more beautiful. Hence the term 'belladonna', which means beautiful lady.

Talking of which, Eve, the beautiful lady with whom Adam shares his life, is still asleep on the other side of the bed. Her long dark hair is cascading across the pillow, and the almost imperceptible rise and fall of her shoulders is the only visible sign that she is breathing. By now, Adam's own pupils have adjusted to the light sufficiently for him to see roughly where he is going when he gets out of bed. As he looks down at his mobile phone, his eyes converge on the slim keyboard and focus on the tiny digits on the screen. It tells him it's now 6.31am. He is grateful for the wake-up call, but equally aware that this wretched little device is going to be a constant source of stress throughout the day, as various people in his working and domestic life make demands upon him. But he can worry about that later. It's time to go to the ensuite bathroom to empty his bladder and get himself sorted for the start of the day. To do that, he has to stand up.

The simple act of getting out of bed can be hard for lots of people. It is warm and cosy in bed. You feel secure and relaxed. You are still a little sleepy. Given the chance, you could easily drift back into some sweet, hedonistic reverie. What is more, it is colder in the room than in bed, and that whole ritual of washing and dressing in the morning is so tedious. Mechanically, however, the physical process of getting out of bed should be simple and straightforward.

Children have no problem at all getting out of bed in the morning. They're usually the ones who are up at six o'clock, throwing off the covers, running into your bedroom and jumping on you in a frantic effort to start the day off with a bang. Unlike adults, most children actually want to get out of bed in the morning, but they still have to perform the same physical movements in order to do so.

At whatever age, getting out of bed involves throwing back the covers, sitting up, bending your knees, swinging your legs over the side of the bed, pushing up with your arms and shoulders, leaning further forward to bring your weight over your feet and then straightening your legs. Each separate action calls into play many specific muscles, including the rectus abdominis, transversus, erector spinae, psoas, quadriceps and gluteus maximus, to name but a few. The contraction and relaxation phase of each muscle has to be timed to perfection so that no awkward jerking or robotic clumsiness occurs. It is this smooth coordination of multiple muscle groups that makes fluent and effortless movement of the body possible and enables you, for example, to get out of bed while holding a full mug of tea without spilling it.

Just rising in the morning is both a partly conscious and a partly unconscious act, involving two different types of nerves. Nerves which instruct muscles to contract and move our bodies are called motor nerves; other nerves which respond to sensations from the external environment are called sensory nerves. The part of your brain called the primary motor cortex is responsible for sending the commands to your muscles that will make them do the job you want them to do, and they will be sent via your spinal cord and all the various motor nerves which emerge from it. While these conscious commands originate from higher frontal brain regions including the premotor and supplementary motor cortices, the unconscious bit of getting out of bed (the bit that you don't even have to think about) first begins in a different part of the brain called the parietal lobe. You might think that getting out of bed means making a decision in your mind and then acting on it to execute the movement. However, it is more complicated than that. In fact, the unconscious part of your brain prepares and begins to initiate movements about two seconds before you consciously 'decide' to do them. The conscious decision itself kicks in only a fraction of a second before you do it. So whatever our conscious mind is telling us to do next, it is really only recognizing instructions from our unconscious mind.

Adam has decided that any moment now, much as he doesn't want to, he is going to get out of bed. That's his Plan A anyway.

If he was one of the 150,000 people in the UK every year who suffer a stroke, it might take him 20 minutes or more, or even prove impossible on his own, but it's an automatic action he has performed every day since the age of 11 months and it's as natural and unconscious as breathing itself. First he has to coordinate his vision.

There are six external ocular muscles which operate the eyeball. They are attached to the bony eye socket and to the tough, fibrous, white outer coating of the eye called the sclera. There is one muscle to the north, south, east and west and two more oblique muscles attached at an angle, which help to rotate the eye so that now not only can he look up, down, left and right but he can smoothly turn his eye through an entire and fluent circle of vision. Three cranial nerves are also involved, the oculomotor, trochlear and abducens, any one of which if damaged will produce obvious abnormalities of eye movement and double vision when looking in certain directions.

None of this of course applies to Adam, as he is stumbling and groping his way towards the bathroom. The only time he has had double vision recently was when he had a few too many drinks at the office Christmas party and had to be sent home in a taxi. Right now, he is OK. He has successfully navigated his way to the bathroom. There in front of him in sharp focus is the shiny, welcoming face of the toilet bowl.

Suddenly he knows he has to pee. It's an instant visual cue that he simply cannot ignore. His bladder has been full for a while now, with all messages about its need to be emptied completely suppressed by his unconscious state while he slept. Now, however, the strong urge to empty his bladder cannot be delayed any longer. The nerve signals relayed with increasing urgency up his spinal cord to his brain are registered and returned as express orders from above to simultaneously contract the detrusor muscles in the bladder wall, and relax the muscles in the purse-string-like sphincter at the base of his bladder. The result? A satisfying and discomfort-relieving gush of urine with an expert trajectory perfectly bisecting the water in the depths of the toilet.

Why do the males of the species take such a pride in that? Why is it such an enduring source of amusement? Why watch the yellow waterfall plunge with a roar into the foaming pool below as the bubbles rise and the colour darkens, then play with the jet from the top left-hand corner of the water reservoir to the top right and then all around the bowl in a methodical clockwise direction, only going off track to sink the remnant of loo paper your partner failed to flush some time during the night? They are questions to which women will understandably never want to know the answer. But if they ever did ask, men could tell them. It's fun. It's the first little game of the day. And because they can. Unless, of course, they have shy bladder syndrome.

SBS or paruresis is a kind of phobia where the sufferer finds it difficult or impossible to urinate in the presence of other people, such as in a public loo at the office or in a busy urinal at a football game. This holds true even if you are completely intoxicated on several pints of beer and your bladder is full to bursting. Women can be affected too, but it is much more common in men and usually occurs just because that tiresome little sphincter adopts a mind of its own and stubbornly refuses to open. A high level of self-consciousness can completely inhibit the nerve impulses from the brain that are screaming, 'Relax, you stubborn little sphincter, I'm in agony here.' That asinine muscular ring remains closed tighter than a dodgy mussel with rigor mortis.

For some, it can be a brief occasional episode in situations where other people are very close by. For others, it can totally prevent them peeing in moving vehicles such as in train or aeroplane toilets. If the sound of their peeing can be over-heard, it can stop them emptying their bladder even when they have an entire public loo to themselves. In severe cases it may mean that they can only urinate when alone at home or when they use a catheter to bypass the function of the sphincter and empty the bladder mechanically. Shyness and embarrassment is part of the problem, and it often begins during one's teenage years when self-consciousness is at its height. But in extreme cases it really can affect the quality of a sufferer's life, restricting travel and getting in the way of forming intimate relationships. Maybe that is the reason why other men who are not affected by it feel the need to celebrate in such an exuberant manner when they let fly into the toilet bowl. Interesting. One of those men's problems Adam is glad he doesn't have.

As he looks up, the message from Eve is plain to see. She's clearly fed up with having to cope with the splashes that the men in the household have made around the toilet seat and on the floor. A mass-produced mini-poster balanced on top of the cistern comes straight from the Management, saying WE AIM TO PLEASE. YOU AIM TOO, PLEASE. Adam immediately assumes it is intended for their teenage son, not himself, and he smiles wryly at the perfection of his own bathroom etiquette. His stream was good. His aim was true. No suggestion of an enlarging prostate just yet. Urine colour was a bit darker than usual though. Should he be worried? What's normal? Does it change much from day to day? Is it a sign of disease? Is he being a hypochondriac? 'Come off it,' Eve would say, 'he's a man.' Of course he is a hypochondriac. It comes with the territory.

Urine in fact can be a colour-coded barometer of many different conditions and diseases in the body, quite apart from the volume, odour and constituents of the urine voided. Urine is around 95 per cent water and 5 per cent degraded cells, proteins that were surplus to requirements, salt and other minerals. And contrary to popular belief, freshly passed urine really is fairly sterile – cleaner than Adam's hands after he has only just washed them and much more aseptic than the egg he fancies poaching and having on toast for breakfast. Some people apparently even like to drink their own urine, believing it to have health-giving properties. Mahatma Gandhi was said to begin every day with a refreshing glass of the stuff. And the first sample passed in the morning was the number-one choice with which gladiators in Ancient Rome would gargle and brush their teeth. Maybe it didn't matter if they were just about to be eaten by a lion.

In the last 24 hours Adam's two kidneys have quietly filtered about 44 gallons of his blood, the equivalent of around two bathtubs full. What he has just excreted is the remainder of the 2 pints or so of urine that his kidneys have produced during that time. It doesn't matter what he eats or drinks, how thick and creamy it is, or how deeply coloured. His kidneys will filter it all, and always produce that watery and yellow-tinged end product. Usually it is the colour of pale straw, but if he is dehydrated or hungover it will be deeply yellow or even dark brown. The yellow comes from the pigment urobilin, derived from the breakdown of the iron-carrying pigment in red blood cells, haemoglobin. The life span of the average red blood cell is about 120 days, so all of them eventually die and are broken down by the liver. Most of the pigment is stored in the gall bladder, digested further in the intestine and then excreted in our faeces, giving them their characteristic colour. Some of it remains in our bloodstream, however, where it is filtered by our kidneys. There, converted into urobilin, it tints our urine that familiar yellow colour. In fact the same yellow chemicals would produce the yellow of our bruises and the brownish hue of the whites of our eyes should we become jaundiced.

Before all this was known, certain alchemists during the

Middle Ages believed that the yellow colour of urine came from gold itself. Not surprisingly this led to frustrating, totally unsuccessful and often quite disgusting efforts to extract that gold. Equally bizarre, it wasn't all that long ago that doctors would test people for diabetes by actually sampling the urine to see if it tasted sweet due to the presence of sugar. That's where the words diabetes mellitus (syphon and honey) come from, because people who are diabetic classically pass large amounts of sweet-tasting urine. These days, doctors dunk chemical reagents on plastic testing strips into the urine samples to detect glucose and are therefore spared the tasting.

Urine isn't always straw-coloured or brown, however. Nor is it always watery or sugary as in diabetics. It can smell fishy and malodorous if infection of the bladder or kidneys is present. It can smell just like asparagus or penicillin if you have consumed these things, as their molecules are small enough to slip through the filtering system of your kidneys. Other medication and foodstuffs contain dyes which can discolour the urine green or blue, and many a hypo-chondriacal man has run screaming to the doctor's with bright red urine only to be told that it's due to the beetroot he's eaten. Urine can be browny-black as a result of jaundice or milky white and cloudy if phosphates are present after a large protein-rich meal. It can be frothy if bile is contaminating it, and long thin threads may be visible in it if the urethra, the water pipe, is inflamed or infected. You can tell a lot about a person from a good inspection of their urine. As far as doctors looking for a diagnosis are concerned, it's almost too precious to flush. But flush it Adam will because it's time to switch on that light above the vanity mirror and see whether he is looking as bad as he actually feels today.

There's so much you can tell about yourself from your eyes. And, for that matter, about other people's health. Bags under the eyes are a dead giveaway, a sure sign of tiredness or of burning the candle at both ends. The skin around the eyes is much thinner than elsewhere in the body, and with advancing age and under the influence of hereditary factors, puffiness can become more prominent or even permanent in some people. This is because the skin gradually loses elasticity and begins to sag. In some folk, the fat that cushions the eyeball pushes through weakened muscles and herniates into the eye bags. Others on medication such as steroids and those who smoke can lose some of the supporting tissue called collagen under the eyes, which keeps them looking young and fresh. Eye fatigue and irritation is another factor that can make eyes look puffy. Heart disease, thyroid and kidney problems may also increase fluid retention and the force of gravity inevitably exerts an increasingly negative effect as time goes by. Some people look like they should have had blepharoplasty (eyelid surgery) years ago. Most people, however, can't afford it as it is only available privately and costs several thousand pounds a throw. You see people on the train or the bus and you look at the areas surrounding their eyes and you can see the early signs of xanthelasma, the raised waxy yellow plaques indicative of high cholesterol levels, and a visible risk factor for future heart disease. If the white of their eye has turned yellowy-brown, they are jaundiced and it's a sure sign of liver or blood disease. If you get really close to other people, as you sometimes have to on the underground in rush hour, you can see at the outer edge of their iris the cream-coloured ring known as arcus senilis, again a sign of hardening of the arteries, high cholesterol and increasing age.

If this close examination of other people's eyes isn't enough to make you blink, the wind and the dust blowing along the pavement or the railway platform certainly will. And your eyes blink constantly. Your eyelids are acting as shutters, opening and closing repeatedly to stop foreign material from entering the eyes. Your lids feel parchment-thin and delicate, and they are, but surprisingly humans have some of the thickest eyelids in the whole of the animal king-dom. You'd think that elephants had much thicker eyelids than ours, but in fact theirs are slightly thinner, and just look thicker because they're very wrinkly and folded over in festoons. Their eyelashes are certainly thicker and longer, but they can't compete with the thickness of our lids. The largest whales also possess thinner eyelids than we do, even thinner than the eyelids of the common mouse. Meanwhile, octopuses and squids have no eyelids at all. Despite having very large eyes in proportion to their overall size they can retract them into their bodies, enabling fat to bulge in front of the eye for protection. So they don't really need eyelids. Cosmetically it's not an attractive look, and most of us would be unwilling to trade in our thicker eyelids despite the fact that they can irritate like hell.

To allow our lids to blink and move smoothly over the

front of our eyes without friction, tears produced in the lachrymal glands in the outer part of our upper lids flow across the eye to lubricate it and wash away any potentially harmful material such as chemicals, allergens and dust. They are salty and contain a natural antiseptic, which inhibits the development of an infection or inflammation. The tears move across the front of the eye to the inner part of the lower eyelid where they normally drain away through the tiny punctum, which is just visible to the naked eye at the edge of the lid. From here they run down the naso-lachrymal duct into the nasal cavity, which will be no surprise to anyone who finds that whenever they cry, their nose runs. Blinking certainly helps this mechanism, without which tears would spill over our lower eyelids on to our faces – a tiresome but common condition that goes by the lovely name of epiphora. Sometimes the lower eyelid can sag and fall away from the surface of the eye, a condition known as ectropion, preventing tears from draining away through the naso-lachrymal duct. They therefore run down the face instead, especially in cold windy conditions. Fortunately ectropion can be treated surgically.

For most people, blinking occurs a dozen times a minute, a totally involuntary response of which we only become conscious if our eyes are dry, tired or sore. Every day, during our waking lives, we blink over 10,000 times. Our lids never become tired or heavy in normal circumstances, but when a tiny fleck of dust or foreign matter lodges under the lid, it's a different story. It scrapes repeatedly against the sensitive cornea at the front of the eye, making our eyes red and giving us a headache. When this happens, the first aid book tells us to use sterile water and a specially designed eye bath to flush away the debris. But how many of us actually possess an eye bath? How many of us can really face a three- or four-hour wait in Casualty just to have the nurse remove the foreign body with a cotton-bud tip?

If this should ever happen to you and you're standing as Adam is now, hovering over the sink in his ensuite, it's just as easy, if not hygienically ideal, to fill the sink with water at body temperature, plunge your face in, open your eyes and have a good look around the bowl. This is usually sufficient to dislodge the fleck of dust and take away the attendant discomfort.

But why is it, you ask yourself when you open your eyes underwater, either in the sink or in the swimming pool, that everything is blurred? It's OK if you're wearing swimming goggles, where the view remains crystal clear, but without them everything is hazy. Any optician will tell you this is to do with refraction indexes. Air has a refraction index of 1 and water has a refraction index of about 1.33. The lens in the human eye is used to focusing light coming through the refraction index of air, not water, so the water acts a little like a contact lens, but one which hampers our vision rather than improving it. Put the swimming goggles on again so that the layer of air in front of our cornea is restored, and hey presto, we can see clearly again.

Yes, eyes are incredible organs, Adam tells himself, as he switches the light off above the vanity mirror and prepares for the next task of the day. It's just a pity that he has to rely increasingly on glasses to see small print and that eyestrain is becoming a regular nuisance. This very day, however, he will be seeing Professor Reinstein to get the long-overdue check-up that will tell him whether laser blended vision could relieve him from the purgatory of having to wear glasses or contact lenses.

Once more he blinks at his reflection in the bathroom mirror. But he realizes he has been staring in the mirror for far too long, now, and if his wife or children catch him they'll accuse him of being vain. As if!

Chapter 2

WHY IS IT, Eve asks herself, that the man she chooses to spend her life with has to make so much noise in the morning? His mobile phone is resolutely set on loud, even though he's the only one getting up at 6.30, and it seems to take him an age crashing and banging around on his bedside table to turn the wretched thing off. He charges around the bedroom like a rogue elephant, loudly throws open the bathroom door and then all Eve can hear is Niagara Falls as he relieves himself while passing wind at the same time. Why do men do that, she wonders? Why aren't they capable of peeing without farting? And if they aren't, why can't they at least close the door? But now it's gone quiet again. So it's a fair bet he's vainly studying his face in the mirror. Just like he does every morning, regular as clockwork. Why is it us humans are such creatures of habit? Especially in the mornings when we get up?

Talking of which, it's nearly time for Eve herself to get up and sort out the kids. But she cannot help lying there a little longer, listening to the birds cheeping and chirping so happily outside. That is, when they are not interrupted by a lorry going past, a plane flying overhead and, for that matter, the distant stream of Adele's 'Someone Like You' floating up from next door. That'll be Michelle in her bedroom with the window wide open as usual. Eve can even hear her singing along. Bit flat, if she's to be honest. She'll never make it on *Britain's Got Talent*, Eve is thinking. Isn't it incredible what you can hear when you really listen? Birds, traffic, planes, music, voices, footsteps, doors, running water or even your partner peeing. And you can detect the pitch, tone, loudness and direction of all of these sounds at any time. How do you do that? And what would it be like if you couldn't? If you were profoundly or totally deaf? What would it be like not to hear? Unable to listen?

For a start, you would miss out on all that sensory information about your immediate surroundings. And you wouldn't be able to communicate through speech or music. But how would you even think internally? If most people like you think in words inside their head, how would you think if you'd never ever heard words? It's a question which scientists have been trying to fathom for years. What they do agree on is how people who can hear actually manage it. And it's amazing.

The sounds that we hear are basically vibrations of the molecules in the air around us. The number of vibrations every second is the frequency, and this determines how high or low any noise sounds – the pitch, in other words. Pitch is measured in units called hertz (hz), and the higher the frequency the higher the pitch and vice versa. The loudness of the sound, or intensity, depends on its power and is measured in units called decibels (dB). Every time the power of the sound increases by 10dB, your ears hear double the loudness – so noises you hear at 40dB, such as the background sounds to a quiet walk in the country, will appear twice as loud at 50dB. So how loud is, say, the ticking of your lovely new Rolex? About 20 dB. Normal conversation? Around 60dB. Eve's friend Tanya at a party? At least 90dB, no problem. That's 10dB louder than general traffic.

Adam's pretty loud too. Like his shirts. He's trying to be Pavarotti in the shower right now but the acoustics aren't great in there, and quite frankly, as far as Eve's concerned, it's more a quality issue than a volume one anyway. Adam's quite a loud snorer too and Eve has heard that some people can snore as loudly as a motorbike backfiring or a jet engine in action. Could that be possible? Probably not. Not quite anyway. Your average motorbike will register 85–90dB and a jet airliner at take-off a whopping 150dB. That's loud. And it's one of the reasons why so many people in the aviation industry, along with those who shoot guns or work with munitions of any kind, often used to suffer permanent acoustic trauma and become deaf. These days, they wear ear defenders to block out the loud sounds.

Now it's the kids who are more likely to damage their ears and suffer from future hearing loss as a result of playing their personal stereos too loud and dancing in front of the speakers at rock concerts. Black Eyed Peas at 100dB? Bad enough, Eve thinks. Any rap artist at 110dB? Insufferable. Eve can't help thinking that maybe in that situation going deaf would be a welcome consolation. Her children of course wouldn't agree. They seem to detect some kind of melody in it. And some sort of meaning they can relate to in the tortured, explicit lyrics. 'Sorry, rappers,' says Eve to herself. 'I just don't get it.'

Ben's the worst. Eve's son has forever got his earphones stuck in his ears, so obviously he never replies when she asks him a question.

'What, Mum? I didn't hear you.'

'Well, take your earphones out of your ears, then. Do you need money for lunch?'

Ah, right. He's heard her now. It must be that selective hearing thing that only teenagers and husbands suffer from. What was it she told Ben recently? Wearing headphones for just one hour a day can create a perfect habitat for bacteria. Warm, moist and stagnant, similar in fact to the habitat of a swamp. In those 60 minutes, the number of bacteria actually increases by 700 times. No wonder ear infections are so frequent these days. And, she told him, sitting in front of the speakers at that rock concert last week would have started to damage his hearing in just seven and a half short minutes. Permanently, too. He still did it anyway.

'Doesn't it bother you?' she had yelled at him. 'Don't you care?'

'What?' was all he'd said. She still doesn't know if it was a joke. Talk about pearls of wisdom falling on deaf ears. And now, to make matters worse, Adam has started singing operatic arias.

Comfortably ensconced in the shower cubicle, Adam is allowing the hot jets of water to play over his head and then on to his face as he lifts it up to meet the torrent from above. It's so refreshing.

Why is it there's nothing as effective as a shower first thing in the morning to get you going? he is thinking. When he first turned the shower on, he stood well away from the head because that initial jet of water is always going to be icy cold. And then, bizarrely, as he stepped into that lovely cascade of hot water, set at just the right temperature for maximum comfort, the bits of his skin that hit the water last suddenly went all goosepimply. As if he was freezing. How did that happen? Why was he getting goosebumps from hot water instead of cold? It is a good question.

The skin plays a very important role in the regulation of body temperature. Unlike other animals, humans like Adam don't have a thick furry coat to protect them from the cold. Instead, Adam has a sophisticated system for thermo-regulation governed by a network of special blood vessels in his skin, which are dedicated entirely to maintaining or losing heat according to the circumstances. There's also a layer of fat beneath the skin, the subcutaneous fat, which provides extra insulation as well as skin suppleness and padding. When his body is too hot, blood vessels open up to allow more blood to reach the skin, where it can cool a little.

That's why his skin is looking redder right now, especially over his shoulders and chest, which are exposed to the full force of the hot water. At the same time, his sweat glands are producing more sweat, which is intended to evaporate when exposed to air, to have a further cooling effect on the skin. It isn't able to achieve that right now, as evaporation can't happen while Adam is immersed in the shower. But as his body warms up and he steps out of the shower to towel off, the sweating might become noticeable. Annoyingly, that will probably happen just after he has got dressed. It's always the case when he has been to the gym, had a good workout followed by a short shower and then got dressed quickly because he is in a hurry. But first thing in the morning he doesn't want to be drenched in sweat before he has even left the house, so he needs to turn the shower temperature down.

The goosebumps he noticed when he first stepped into the shower occurred because his body recognized a big contrast between the heat of the water and the relatively colder air on his dry bits. His brain interpreted this as his entire body being cold and his skin reacted accordingly. When his body is too cold, those same blood vessels that cause his skin to redden when hot constrict. Simultaneously the tiny muscles attached to each individual hair on his body, the erector pili muscles, contract, lifting the hairs up vertically and pulling on the skin to produce goosepimples. In evolutionary terms, the hairs were designed to trap an insulating layer of warm air close to the skin, similar to the effect of double glazing or wearing a wetsuit. But with the sparse hairs most humans have on their body these days, the effectiveness of this process is minimal. So Adam's skin reacts to the outside temperature all the time in an effort to keep his body temperature the same. That's part of homeostasis, maintaining a constant internal environment for his body. But how does his skin know what the exact temperature of the water is? If the blood vessels and erector pili muscles act as your skin's thermostat, how does his skin judge how hot or cold anything it touches actually is? Where is its thermometer?

The thermometer is in fact made up of millions of free nerve endings in our skin called thermo-receptors. These have properties that enable them to respond to changes in temperature. There are hot and cold thermo-receptors that respond to a whole range of temperatures, overlapping to a large extent. Together they provide information over a spectrum of skin temperatures. As they react, these thermo-receptors generate electrical activity in sensory nerves which is forwarded to the spinal cord and upwards to the brain. There, it is integrated, coordinated and, if significant, inter-preted as a conscious perception of something our body needs to be aware of. If the shower tap has been left on its cold setting and the water is freezing, Adam would certainly need to be aware of that. Thanks to all those sensitive little thermo-receptors, Adam gingerly touches that icy jet of water at the beginning of his shower and can then avoid an unpleasant experience. If he had stepped into the shower straight away, he says to himself, that's when he really would have got goosebumps. And stinging pain in his skin. And, heaven forbid, there would have been definite and somewhat alarming alterations in the size of his manhood.

Why is that? Why does a man's penis shrink right down when it is cold and only attain its normal size when it's warm? And why does his scrotum scrunch right up and push his testicles up inside his groin as if he is a sumo wrestler about to do battle? As far as his body is concerned, he *is* just about to do battle. With the elements.

In anatomical terms, his penis is an end-organ without collaterals. A collateral system of blood supply, that is. This means that because it is an extremity of his body, like a finger or toe or ear, the blood circulation flows to it and back again, but not through it to the other side. There is no other side like there is, say, in the stomach or liver or brain. Blood goes along it to the end and back again. This makes his penis, like all his extremities, extremely vulnerable to the cold. When he is cold, his circulation shuts down to preserve body heat. At extreme levels of cold, it shuts down so much that no blood flow gets there at all. That's why fingers, ears and toes are so susceptible to frostbite. Ask any mountaineer or explorer, especially those who developed gangrene as a result of frostbite and had to have bits of their extremities amputated (if they didn't fall off by themselves). Sacrificing the extremities, alarming as it sounds, is simply the body's way of preserving heat for more important organs such as the brain, kidneys and heart. In physiological terms, at least, these facts are irrefutable. But Adam remains unconvinced. What organ could possibly be more important than his todger?

But he need not worry. His precious private parts are not about to drop off on to the shower tray just yet. They are, however, retracting alarmingly quickly towards his lower abdomen. In fact, they are almost invaginating. For the same reason that blood vessels in his skin contract to conserve heat, the blood in the spongy erectile tissue within the shaft of his penis is emptying back into the general circulation of his body, leaving him somewhat more flaccid than he was on his honeymoon.

As for his scrotum, the power of the contraction of the dartos muscle, which covers it, shouldn't be underestimated. It has changed a sac the size of a conference pear into something no bigger than a lychee, and more or less the same colour. Why? His scrotum isn't an extremity, like his penis. The reason his scrotum contracts so tightly when he is cold is different. Normally, sperm production is optimal at about 2 degrees Fahrenheit lower than core body temperature. That's about 95 degrees Fahrenheit. This explains why a man's testicles are located outside his body. In the male embryo they actually develop high up in the abdomen next to the kidneys and migrate downwards as the pregnancy becomes more advanced. At birth, in a male baby, they drop down into the scrotum where they lie, totally innocent and unassuming, until the sex hormones kick in at puberty and turn them into exquisitely sensitive testosterone-producing plum-sized sperm factories. And this is the thing. A man's body intuitively knows that he is totally dependent on them if he ever wishes to mate with a member of the opposite sex to create a family and pass down his genes to future generations. So his natural instinct is to preserve these family jewels at all costs. Quietly producing sperm all day at a cool 95 degrees Fahrenheit is all very well, but just like any other tissues of the body, the testicles start to baulk at any temperature below that. So his body invites them back inside, back in from the cold.

The sperm-carrying tubes – the vas deferens – and the veins, arteries and nerves which are attached to the testicles are all packed together in a neat little bundle called the spermatic cord. It runs downwards along the inguinal canal (a narrow tube-like structure lying just over the crease where the top of your thigh meets your lower abdomen) and suspends the testicles in the corrugated bag of hairy skin that is the scrotum. It also carries muscle fibres within its bundle known as the cremaster muscle.

This is a fun little muscle because it possesses a reflex. A bit like the knee-jerk reflex, only nicer. Whenever a man's inner thigh is stroked upwards (preferably by someone other than himself) the testicle on the same side noticeably moves upwards. He will feel it too. This is the cremasteric reflex, and its evolutionary function is to remove the testicles from harm if something nasty is attacking your legs. Like a shark, for example, or a hungry lion who thinks meatballs might just be on the menu. But the reflex also plays a role when you're very cold. To bring the testicles back inside the rest of the body for warmth and to preserve your ability to father children in the future, the cremaster muscle tugs them up inside the inguinal canal where you can feel them, hard as a walnut, just above the crease at the top of your thigh.

It's a useful if somewhat unusual protective function which professional sumo wrestlers in Japan have made into something of an art form. They can manipulate their testicles at will, so far along their inguinal canal that they actually pass through an internal ring of muscle in the lower abdominal wall, to slip wholly inside the abdominal cavity, where they can no longer be felt or injured. That is of course why they do it: to protect their sensitive testicles from physical harm. Any man knows how crippling a kick in this area can be. Most canny women know it as well. Some have occasionally had recourse to such a kick to discourage unwelcome advances or assaults. So you can kick a sumo wrestler wherever and however often you like and all they will do is smile at you. Before breaking your neck. No opponent can get the upper hand with a blow where it usually hurts most. It makes the contest much more equal.

Adam's thoughts are interrupted by Eve, who's shouting at him from the bedroom.

'Are you fiddling with your bits as usual?' she's asking.

'Actually I'm just wondering what it means when my skin gets all wrinkly after I've been in the shower for a while.'

'It means you've been in there far too long and you've probably left no hot water for anyone else.'

But Adam was just about to turn the shower off anyway. Why is his skin so wrinkly? Look at his fingertips. The ends of his toes. All waterlogged and crinkly. What's happened there?

What's happened is that the outer layer of his skin, which is dead, has absorbed more water than the inner layer of living skin beneath it. Since this is impervious to water from the outside and it can't get any bigger, the outer layer has to corrugate and pleat. In addition to that, the lower layer of his skin actually starts to lose some of its natural protective oils through sweating and the emulsifying effect of the detergent in the soap Adam is using, and begins to leak water from inside the body. So when he immerses himself in water for any length of time, his body cannot absorb water and swell up like a sponge, but it can leak water to the outside, leaving him looking like a wrinkled prune. Talking of which, it occurs to Adam that the social visit to Eve's elderly parents tonight is somewhat overdue, and has been written in the diary for some time. Eve's been on about it for days now.

Back in the bedroom, Eve is doing her best to ignore Adam's noisy ablutions. Everything seems so loud these days. Anyone like her, who's worked in the radio, TV or film industry, will know you can hardly go anywhere in the world to record a few minutes of conversation without some kind of loud interruption to the soundtrack. Noise pollution can be a real issue. There will always be the beep beep beep of a reversing lorry, a car horn tooting, a police siren blaring, a chainsaw revving or an electric drill at full power. It's always Sod's Law too that the loudest intrusion will always start up just before the longest, best and very last take is about to finish. Either that or everyone in the media is paranoid.

Suddenly it's peaceful again. Adam's out of the shower, drying off. For a few moments at least, the only background sound is birdsong. It's so calming. But how do we actually detect those different sounds and tell them apart? It hasn't got much to do with that floppy pair of flesh funnels sticking out of the sides of our heads that we call ears, that's for sure. These are a particularly handy place from which to hang those lovely Tiffany earrings Adam gave Eve for her last birthday. But as a parabolic satellite dish collecting sounds from her environment and funnelling them precisely into that hole in her head called the external auditory canal, they are fairly rudimentary.

Despite this, sounds do make it into the canal, travelling down past all the earwax and skin flakes until they reach the delicate cellophane-thin parchment suspended across it known as the eardrum. Only 0.7 inches or 17.5mm in diameter and 0.05mm thick, it is indeed comparable to a drum, because it is so tightly stretched across the width of the canal, where it will resonate clearly as it is struck by sound waves despite moving less than a billionth of an inch as it does so. That's what you call sensitive. And that's why perforating your eardrum through a blow to the head, diving underwater or just doing something as daft as shoving a cotton bud, matchstick or hairpin too far up your ear canal can seriously affect your hearing. Only the other day Eve had read about a security guard who had perforated his own eardrum with his walkie-talkie aerial while he was driving. Serves him right, Eve had thought. What a foolish time to try to remove earwax.

What exactly is earwax anyway? she'd asked herself. What does it consist of? Cerumen, to use its real name, is basically a mixture of watery secretions from modified sweat glands and more treacly secretions from oil-producing sebaceous glands. Sixty per cent is keratin – the material which makes toenails and fingernails hard – but it also contains fatty acids, cholesterol and dead skin cells. A cleaning agent with lubricating and antibacterial properties, cerumen scours the ear canal and protects it. The epithelium – the surface layer

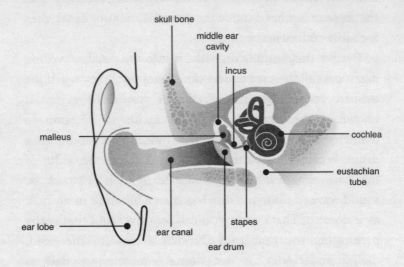

The Ear

of skin inside the ear canal – gradually migrates outwards from the eardrum, acting like a conveyor belt transporting dirt, germs and dust out of the ear. Starting as slowly as fingernail growth and picking up speed as it reaches the exit of the ear, it also collects earwax produced in the outer third of the canal and carries this gloopy mixture of waste away to the outside.

On the other side of the eardrum lies the middle ear, an air-filled cavity no bigger than a peanut. This contains three tiny bones, the malleus, incus and stapes, which translated from the Latin mean hammer, anvil and stirrup on account of their shapes. They're the smallest bones in your entire body, all three easily capable of fitting together on top of a 1p coin. Attached to the stapes bone is the smallest muscle in the body, the stapedius, which is only one-twentieth of an inch long, half the size of the stapes bone itself. Its other end is fixed to the ceiling of the middle ear cavity, and it can suddenly contract to dampen down any very large sound vibrations, protecting your ears from loud noises which might otherwise be painful or lead to permanent hearing loss. Connecting closely to one another, the three fragile middle ear bones transmit vibrations from the eardrum to another membrane at an oval window situated at the entrance to the inner ear, the mysterious and convoluted maze of passageways and fluid-filled channels going off in all directions and buried deep within the mastoid, the solid bit of skull bone directly behind the ear canals. Here, inside this labyrinth, is the cochlea, containing the sensory receptor for hearing, as well as additional anatomical architecture designed to give us balance and equilibrium. When those birds sing, or our alarm goes off, when our favourite song is played on the radio or our lover whispers sweet nothings in our ear, sound waves are transmitted through the eardrum and the middle ear ossicles to the oval window, where the membrane there forwards the vibrations to the sensory receptor for hearing.

The receptor, which is also known as the spiral organ of Corti, runs through the inside of the cochlea. It consists of thousands of sensory hair cells attached to a membrane. Tiny sensory hairs emerge from each sensory hair cell and pass through a second gel-like membrane above. When the fluid in your cochlea is moving, the first membrane vibrates and pushes the sensory hairs against the second membrane. The movement of the sensory hairs is then converted into nerve impulses which travel along the cochlea nerve to your brain.

So how does your brain interpret and make sense of different vibrations? One of the membranes in the cochlea detects the component frequencies of incoming sound. This membrane is particularly good at detecting frequencies or tones. It's flexible and vibrates when sound reaches it, but it doesn't vibrate to the same extent all over. One end of it vibrates most at low frequency tones, and the other end at high frequency tones. This gives the membrane tonotopic organization, in other words organization by tone, like a keyboard instrument such as a xylophone. Just as you can learn which end of the keyboard is being used by recognizing different tones, your brain can learn which frequencies are being picked up by which part of the cochlea membrane so that it can understand the type of sound. So your inner ear is really your body's inbuilt microphone, and it enables your brain to detect different frequencies and tones and tell the difference between Gary Barlow and Michael Bublé. That's the function of those 20,000 hair cells and the membrane attached to them.

But if that's how we physically hear and can discern the quantity, the quality and volume of sounds, how do we know where they're coming from? How can you locate the source of sounds? The answer is easy: you are born with two ears, so you have stereoscopic sound capability from birth. If a sound comes from your right side, for example, it will reach your right ear just before it reaches your left ear. It will also be slightly louder in your right ear and as a result you can recognize the sound as coming from your right. That's basically how you can enjoy the sound of the different birds in your garden and know whether the ice-cream van with its chimes is coming towards you from the right or moving away to your left.

At this particular moment in time, a sound Eve didn't particularly want to hear is coming from directly above her and slightly to her left. For some reason, Ben is threatening to invade Poppy's room and she's screaming the house down for him to stay out. A tiny part of Eve's brain is telling her this, while a totally different part is telling her that it's probably time to get up.

Chapter 3

IS DYSLEXIA A sign of a mis-spelt youth? Ben thought of that line while lying awake at 3am with a busy mind and liked it so much he repeated it in his own head a few times so he could still remember it in the morning. He is thinking he might try it out on Poppy later, to see if she also thinks it's funny. An amusing little *jeu de mots*, he reckons, aware at the same time that it might well be regarded as offensive to anyone else like himself with dyslexia. He knows they would probably not consider it 'furry' or clever. Ha ha. Ben laughs out loud. Surely he can make a joke about it, as he is the poor victim who has to live with it.

Dyslexia is by definition a form of word blindness. Ben's was confirmed by the excellent Helen Arkell Centre in Farnham, and he also, as bad luck would have it, suffers from colour blindness. Ben is of the opinion that he must have done something really bad in a former life, especially as he was told that both conditions are untreatable.

But are they? His biology tutor has told him about a

revolutionary new scientific approach that suggests they might not be. Apparently, he could use special contact lenses to improve things fairly dramatically. But is it even remotely possible that wearing tinted contact lenses could improve both of these untreatable conditions at the same time? He has tried to look at it logically. How is normal colour vision meant to work? What happens at the back of Poppy's eyes that is different to Ben's and means that she can be totally obsessed with the colour of her hair whereas Ben isn't remotely interested? Ben must have missed this part in biology class.

In fact, there are two types of light-sensitive cells in the membrane at the back of the eye (the retina), the rods and the cones. There are up to 120 million rods distributed throughout the retina, and while all of them are sensitive to all visible light, they contain only one type of pigment and can't distinguish colours. So it's these rods that are mainly responsible for vision at night. On the other hand, there are 6.5 million cones, which are capable of providing detailed vision and colour vision. Each one responds to red, green or blue light, although they can only work well when it's bright. They're mostly concentrated in one central part of the retina, known as the fovea. When light strikes the retina, the rods and cones generate electrical signals which stimulate further impulses in the nerve cells to which they're connected. These transmissions travel along the optic nerve to the brain, which in turn interprets their meaning by not just telling us what we see but offering it to us in glorious Technicolor.

But what about Ben and his colour blindness? His colour vision deficiency is caused by an inherited defect of the light-sensitive pigment in the cone cells of the retina at the back of his eyes and, as is usually the case, this affects his ability to pick out greens and reds. In the UK, about 8 per cent of males are affected like Ben, and they retain the defect for life. Up until recently, colour blindness had been considered incurable, something that just had to be lived with. As far as Ben's concerned, it isn't as bad as dyslexia because he has never known anything else. He has never really seen the deep red of a bunch of roses or the lush green of the grass on the cricket pitch in the village. He sees most things in varying shades of grey. Even traffic lights. He can discern the different shades and thus identify the colours, but he just doesn't see them as everyone else does.

Dyslexia, however, can be more of a problem, as Poppy is always reminding him. 'You haven't got dyslexia, you're just thick,' she says. But Ben has got dyslexia. And by way of reply he points out to his sister that at least he hasn't got dyspraxia, dyscalculia, dysmenorrhoea, distended ego, disused brain and general disability like she has. His dyslexia is a specific reading disability which, although it's a bit of an inconvenience to him, has at least been responsive to the specific remedial teaching he has had in the past, when he learned to develop tricks and strategies to overcome his reading difficulties. That was always the conventional wisdom anyway, until doctors came along with the notion that tinted contact lenses could not only significantly improve reading ability in people with dyslexia, but also help people with inherited colour blindness to perceive colours more strongly.

Eve had read about it and taken Ben to Brackley in Oxfordshire to see some very clever people at a company called Cantor and Nissel to find out whether they could help him. They explained that his colour blindness had been inherited through the female side of the family. This is because women have two X chromosomes and men have one X and one Y chromosome. The gene responsible for colour blindness sits on an X chromosome and only manifests itself in males, whose shorter Y chromosome allows the colour blindness gene to express itself because it is unopposed by a normal gene on another equally long X chromosome. For this reason, colour blindness in women is very rare. It was complicated, but Ben got the hang of it in the end. They also explained that Ben's dyslexia might or might not be related to the fact that he was born slightly prematurely, when his brain cells were still whizzing around trying to connect with one another in the right way.

The doctors at Cantor and Nissel had carefully assessed him and manufactured some ChromaGen contact lenses for him to wear. They anticipated that these would not only improve his reading speed and accuracy, but also help him to appreciate colour vision more acutely. After a battery of tests, the ophthalmologist had prescribed him a pink contact lens for his left eye, and an orange for his right eye. Using temporary lenses they'd measured his reading speed and accuracy with and without them, and managed to improve his score in both categories enormously. The whole thing had seemed too good to be true.

Ben had even been treated to a guided tour of the laboratory where they were doing some really cool stuff. There, operatives in long white coats, sterile caps, masks and gloves were making every sort of contact lens imaginable. There were lenses for people with long sight, like Eve, lenses for people with presbyopia, like Adam, lenses for people with astigmatism and whose corneas need to be splinted into a symmetrical shape, and monocular lenses for people who require both long and short sight correction but don't want to rely on glasses. Ben remembered that that was what his dad had been offered, but he had subsequently opted to go down the laser blended vision route. Horses for courses.

There were also weird brightly coloured or patterned ChromaGen lenses for the Hollywood film industry, where they have been used in films like *Star Wars*, *Blade* and *Lord of the Rings*. There were posters of the characters in these epics with their strange costumes and mad-looking eyes adorning the corridor walls, and Ben's doctor had fixed him up with a couple of pairs of fun lenses just in case he wanted to go to any fancy dress parties in the near future. Ben thought these were really cool. One pair had a skull and crossbones on an opaque background and the others were just plain black. Imagine, huge black eyes with no colour around them at all. If the pupil of the eye is the window of the soul, these little lenses were going to be the black hole into the brain.

So Ben had pocketed his interesting little discoveries and then there he was, being driven home in the back of Eve's car, getting used to his new lenses. They had gone over the brow of a hill and a panoramic countryside view had opened up, with a blood-red sun just dropping below the horizon and sending a pink glow throughout the sky. 'Wow,' he remembered saying. 'I've never really seen the colours of the sunset before. That's pretty amazing.' That had been the moment Ben realized ChromaGen really were on to something. It wasn't yet a cure for colour blindness but while he wouldn't admit it to his mum and dad, his schoolwork had become that little bit easier as a result.

Chapter 4

UPSTAIRS IN HER bedroom, Poppy is quickly getting dressed for school. It's bitterly cold outside but because of a ridiculously rigid school policy, she still has to wear their disgusting summer uniform. The dress is red with purple, yellow and blue checks and she is not allowed to wear tights with it, whatever the weather. The obligatory soft flat shoes don't help – hardly Louboutin with killer heels, Poppy thinks – and she is only allowed one pair of earrings and no other jewellery, and certainly no colourful hair bands. If she didn't feel so self-conscious already it might not be so bad, but the uniform makes her feel several times worse. Look at me, she thinks, as she stares in the mirror. Fat. Fat in all the wrong places. Muffin top. Droopy bum. Fried egg boobs. Flabby arms. Orange-peel skin on my thighs, which will look blotchy when I hitch my dress up higher than I am allowed on the way to and from school.

Poppy is Adam and Eve's adopted child and is fifteen going on twenty. Her biological mother, Rawaa, died when

Poppy was just three. Rawaa's parents were Sudanese but she herself was born in England and had been Eve's dearest friend ever since they had attended nursery school together as toddlers. As near neighbours they had followed in each other's footsteps while growing up and, as fate would have it, had come together again after university to work as news and features journalists at the same TV company.

It was in the course of doing what she loved most, campaigning for justice, freedom and the empowerment of Sudanese women, that Rawaa was killed. It was never satisfactorily proven but, during the making of a moving TV documentary about the treatment of women in her native Sudan, she had been murdered by dark forces with vested interests. This left poor little Poppy an orphan.

As Poppy's father was unknown to her and untraceable, it seemed the most natural thing in the world for Eve, as Poppy's godmother and nominated guardian in Rawaa's will, to adopt her as her own daughter.

They already had Ben, who as a five-year-old was very low maintenance and a model child all round, and, as it happened, at the time Eve and Adam had been trying for another baby themselves for over two years without success. Investigations had shown no identifiable cause in either of them and theirs appeared to be one of the 30 per cent of infertility cases where there was nothing specific to treat. In-vitro fertilization and assisted conception were not a route they had chosen to go down at that point. Poppy's adoption therefore felt somehow pre-ordained, almost spiritual. Then, just over ten years later, the time was right for Ruby to be conceived and now, totally unplanned and thanks to a rare failure of the oral contraceptive pill, Eve is expecting again. It's strange how it sometimes works out like that, Eve often thinks. A woman is meant to be at her most fertile before 35 and after 37 her chances of conceiving fall dramatically. For Eve, her fertility seemed to have worked in reverse. Sometimes she wonders if Rawaa is looking down favourably on them from heaven because Poppy's adoption seems to have been the catalyst which sparked it all off.

The adoption hadn't been easy, what with all the extra barriers involved when a child is of different ethnicity. But eventually it was done and they have been a close, loving, solid unit ever since. As far as they are all concerned, Eve is Poppy's mother and Poppy is Eve's daughter. It's never occurred to the other children to think any differently.

Right now, though, Poppy hates her own appearance and she wishes she had more self-esteem and confidence. Adam and Eve are always saying how great she looks, and so are her friends, but she is convinced they're only saying it to cheer her up. She has talked to Eve's sister, Aunt Sally, about it, and she was a real help. Poppy finds it easy and fun to talk to Sally. She always makes her laugh. Poppy can tell her things she can't tell her mum about. Even though, if she needed to, she knows she could tell her mother about anything. Somehow it's just easier to chat to Sally at the moment. Sally hasn't got children of her own yet, which means Poppy can talk to her a bit like a big sister, and that helps her to say what she really feels. The other great thing about Sally is she tells Poppy that she is normal. Poppy feels like she is morbidly obese despite weighing 8 stone 4 when she is 5 foot 8 tall, but Sal says that is 'spot on' for being both healthy *and* attractive to boys. She also says she felt exactly the same when she was Poppy's age. That really helps.

Poppy doesn't understand why she is so moody at times, so fixated on how she looks, so conscious of what she eats and so unable to control and hide her emotions when someone like Ben's friend Frankie pays her attention. And she can't stand the fact that she blushes so obviously. It's such a giveaway, thinks Poppy, and means that everyone deliberately tries to embarrass her, which only makes it worse.

'I used to blush,' Sally told her, 'but I grew out of it.'

Just last year, Sally had taken Poppy out shopping and, over a very long latte and a blueberry muffin, she'd explained about body dysmorphic disorder, the hellish mood swings of puberty and the dangers of anorexia nervosa, without Poppy ever feeling criticized, judged or patronized. She actually had a good laugh about it all, recognizing a lot of the same traits in herself and learning that what she was going through was absolutely normal. She already knew about puberty and the hormonal changes that come with it. But she didn't realize just how many hormones are involved and how complex the whole process is. Nor did she know that hormones like oestrogen don't just affect the lining of the womb to cause periods, but can influence other parts of her body such as her bones, her circulation and her brain. Yes, there are even oestrogen receptors in her brain, which is why women going through the menopause, or puberty for that matter, can experience massive mood swings.

Sally made Poppy feel, for the first time, like it wasn't all her fault. That she wasn't just a gobby cow at heart, as Ben had called her. Sal also told her that she is now producing quite a bit of testosterone as well. She had always thought that was just for the boys. Girls make it too, Sal said, albeit in lesser amounts than boys, because girls also need to grow and develop strong muscles. And they still need to hold their own in an argument with boys, which they usually win hands down anyway. But Sal had also explained that girls don't want too much testosterone either. Otherwise they grow clitorises the size of cucumbers and start speaking like Barry White. Nor do they want the physique of a Russian shot-put champion with biceps like Popeye or to be covered in thick, matted body hair like a gorilla. Sally really made sense of some of the facts Poppy already knew, and she put it all into much better perspective.

And she loved Sal's parting joke about why boyfriends and brothers should always be wary of upsetting a moody girl with premenstrual syndrome and a mobile phone.

'Since you have GPS as well as PMS,' she had explained, 'it means you are a complete bitch and you *will* find them.'

Chapter 5

D AD'S SUCH A prat, Ben is saying to himself. Does he really
think he can sing? Why does he have to wake the whole
house up anyway? Selfish git. It isn't even seven o'clock yet.
So Ben lies back in bed and just yawns. Yes, yawns. A long,
wide-mouthed, achingly refreshing morning yawn.

What's that all about? Why does he do that? What
possible purpose does a yawn serve? Usually people only
yawn when someone else does it first, which has always
puzzled Ben. Can yawning really be contagious? If it is,
what's the point of it? An automatic social bonding ritual to
get him to empathize with his fellow man, or just a futile
exercise in mimicry?

In fact, yawning is a common and universal phenomenon
which humans enjoy from the cradle to the grave. A little
rude in polite company, unless the mouth is covered with the
back of the hand perhaps, but not something worthy of
treatment or psychological avoidance behaviour. Scientists
define it as a 'gaping of the mouth accompanied by a long

inspiration followed by a shorter expiration'. Ben already knows that. But what's it *for*? Is there any physiological use for it, or is it just interesting but a complete waste of time? The answer is, we really don't know. It might help open up the Eustachian tubes that connect the ear passages between the middle ear cavity and the nose to help equalize air pressures and enable us to hear better. It might also help us take on board more oxygen, and get rid of the waste carbon dioxide our lungs produce as a by-product of our metabolism.

Yawning seems to have a place in preventing post-operative respiratory complications too. There is a theory that it keeps our brains more alert when we are bored or drowsy. And yet another theory is that it actually relaxes us and promotes drowsiness and sleep. Scientists seem as confused as everybody else. They have, however, discovered that yawning is closely associated with stretching. So when Ben yawns, he is much more likely to stretch at the same time. Amazingly, people who are paralysed in one or more limbs have been known to carry out stretching movements in these same paralysed limbs when they yawn. It seems brain-damaged people cannot physically separate the two activities, for reasons we really don't understand. Scientists have even looked into whether people who yawn only occasionally compared to others compensate for infrequent yawning with longer-duration yawns. But they don't. So another blind alley then. What is interesting is that yawning does seem to be a normal activity, and definitely related to good health. People who have brain disorders often can't yawn, while individuals who are psychotic apparently yawn less often.

Ben wishes it would make them sing less often too. As far as he is concerned, his dad is psychotic, and he's just started singing another dreadful rendition of 'Nessun Dorma' as he makes his way down the stairs towards the kitchen. It's true what they say about yawning, Ben thinks to himself as he lies back in bed feeling tired, lethargic and with no enthusiasm whatsoever to face the day ahead – it's definitely associated with tedium, boredom and really low levels of arousal. There he goes. He's yawning again now.

Chapter 6

JUST AS EVE is about to get out of bed, in toddles five-year-old Ruby with two vertical streams of green snot pouring from her nose. 'Come in number 11,' Eve mutters before reaching for a tissue and wiping away the stringy muck in one deft pincer movement. She's well practised in the art, having had to do it on a regular basis over several years for her older children. Why oh why do kids this age produce so much snot? Eve wonders. Where does it all come from? Where does it all go? And why do their noses sometimes run like a tap and on other, albeit rare occasions they seem to dry up and behave themselves? Just looking at it makes her want to heave. Especially at the moment since she's 12 weeks pregnant and feeling continually nauseous anyway. It always makes her think of that joke: 'What's the difference between snot and broccoli?' Answer: 'You can't get kids to eat broccoli.' Still funny, she thinks to herself.

Why is it kids always seem quite happy to eat their own snot? Most adults thankfully grow out of that particularly disgusting habit. But some don't. Eve sees them on public transport or stationary in traffic, thinking nobody is watching. Then they shove a finger up their nose, swivel it round a bit to fish out the contents, take a long fond look at it and then, horror upon horrors, lick it off their finger and eat it. Oh yuck. Now Eve does feel sick. How can they do that? Why would they want to? What does it taste like?

What they are doing actually has a medical name, and if there's a name for this Eve reckons there must be a name for almost anything. It's rhinotillexomania. From the Greek *rhin*, meaning nose. And tillexomania meaning the most stomach-turning revolting unsavoury human behaviour on the planet, presumably. Except, of course, the natural saltiness of snot would in fact make it rather savoury. Not as salty as seawater or the frosting on the edge of a margarita glass perhaps, but certainly saltier than, say, lemonade or a comparably green crème de menthe. Only the cognoscenti would know for sure. Eve certainly wouldn't. Thankfully, poor little Ruby's nose is all dry, clean and tidy now. Well, almost, because it's just beginning to trickle down again and Eve's reaching for another tissue.

Children's noses run for a reason. It's basically the same reason adults' noses run, but kids' noses are just that much better and more efficient at it and have a greater need for the protection their mucus provides. Your nose is one of your body's first lines of defence against infection and allergy. Because you constantly breathe through your mouth and nose both are exposed to all of the germs carried around and transmitted by everybody else. Dust, germs and allergens are also drawn into your nose as you inhale, and once they've got past the not very efficient filtering system of your nose hairs, they land on that moist flypaper which is the lining of your conk, called the mucous membrane. Now, because this is quite sticky, it's ideal for trapping viruses and bacteria and smothering the life out of them. It's also covered with millions of even tinier hairs called cilia, which are so small you'd have to lie 25,000 of them end to end to make them stretch just a single inch across. They waft backwards and forwards hundreds of times a minute, some sweeping all snot before them towards your throat, and others gently pushing odour particles upwards towards the olfactory mechanism in the roof of your nose, which is responsible for your sense of smell. The snot that reaches your throat is swallowed and then passes via your gullet into your stomach.

It seems, therefore, that one way or another your snot is destined for your stomach, whether you suffer from rhinotillexomania or not. It's just that in those that do, their snot takes a bit of a diversion. At least in the stomach the snot (or mucus as you should now call it, since it didn't exit the nose from the front), with its pale colour, viruses and payload of assorted bacteria, is neutralized and sterilized by the strong hydrochloric acid produced there. This is probably just as well since it's estimated that your nose produces about 2 pints of mucus every day, and you don't really want to be made too conscious of it. Children though, like Ruby, can sometimes swallow sufficient quantities of infected mucus on an empty stomach to make them feel sick enough to throw up, which they often do anyway at the drop of a hat. Thankfully Eve has learned that saline or 0.5 per cent ephedrine nose drops can effectively disperse that nasal mucus and clear the nose, so she keeps a supply in her fridge at all times.

Bizarrely, in biological terms mucus is actually quite useful. It wraps up potentially harmful micro-organisms in layers of tenacious gunk and gets rid of them. But where does it all come from, and what is it made of, and why the hell does it sometimes turn green? The mucous membrane which lines your nose contains specialized secretory cells called goblet cells which produce a liquid protein known as mucin, made from molecules called mucopolysaccharides. Mucus is made from 95 per cent water and equal amounts of salt and mucin. Mucin is so thick and gloopy it has actually been used to produce certain types of glue. This explains to Eve why scraping the dried stuff off her child's face can sometimes prove so difficult.

On a routine day, the average person is likely to swallow a fair old quantity of mucus without even knowing it. But when you're exposed to a cold virus or to an allergen of some kind such as grass pollen, your goblet cells go into overdrive and start to produce much greater amounts of mucus to eradicate that potentially lethal threat. When it's first produced it's runny and watery, and if it's an allergy that has caused it, it will often stay that way. But if the trigger is an infection, it will gradually turn thicker, creamier or greener as the bacterial waste and the pus that goes with it alters its composition. Finally, once your antibodies have kicked in, recognized the germs for what they are and destroyed them all, no new bacterial waste is formed and the mucus production begins to dry up. Snot becomes thicker and stickier as no new watery outpourings are forthcoming, but the goblet cells cannot simply switch themselves off like a tap, so mucus continues to be produced in smaller and smaller amounts until it finally stops altogether over a number of days or even weeks. That's why, when you have a cold, it often seems to last an eternity. The bugs don't but the mucus does, which is why pharmaceutical companies produce so many expectorants and mucolytic medications in the hope of cashing in on the human race's obsession with snot-free living.

Mucus doesn't just affect the nose, however. You need mucus in lots of different parts of your body to help it function healthily. Just ask anyone who suffers from cystic fibrosis and whose inability to make normal 'unsticky' mucus in their lungs and intestines condemns them to a life of recurrent and often severe respiratory infections, pancreatic dysfunction and weight loss. In your stomach, for example, the tough barrier of mucus which lines its wall protects your body from being digested from within by your acidic gastric juices. You also produce mucus in your middle ear cavity (where it's called catarrh) and in the air passages within your lungs (where it's called phlegm, if and when you ever cough it up). Inhaling cigarette smoke increases mucus production in the lungs, and upper respiratory infections increase mucus production in the middle ear cavity. In each case, the cells lining these hollow passages produce thick glutinous mucus to wrap up any invading micro-organisms or foreign irritant materials and first inactivate and then dispose of them. It's the same principle of self-defence that triggers mucus production in the nose, only the stimulus and location are different.

So Eve can now look at her daughter in a new light. Ruby's snot is a magical thing of wonder and awe. It's protecting her while signalling to Eve, her mother and carer, that her immune system is maturing and reacting in a healthy, predictable way to noxious environmental challenges. Then she sneezes. Without any warning whatsoever. Not even an intake of breath. Not even a throwing back of the head or a little hand over the mouth. *Aaachooo!* And the snot, at least a full half-night's worth that must have been accumulating in the cavernous recesses of her maxillary sinuses beneath her lovely cheekbones, flies in great slimy green ribbons all over Eve before she has had a chance to blink.

'Is breakfast ready yet, Mummy?' asks Ruby as if nothing has happened.

'No, darling,' Eve replies, taking care not to get cross or open her mucus-covered lips too far. 'No, it's snot.'

Chapter 7

A S BEN LIES in bed stifling yet another yawn he knows that the wet-sounding sneeze from downstairs can mean only one thing: Ruby's up. There'll be no more peace in the house this morning. It's time to get ready for college anyway. He has just sniffed the armpit of the shirt he was wearing yesterday. Eeeuuww! It's pretty bad. But he forgot to put it in the laundry for his mum to wash, and that means it'll have to join the other two in the heap in the corner of his room. It also means he hasn't got a fresh one and he'll probably have to borrow one of Dad's, which will be at least one collar size too big and will make him look a bit silly. He curses. Maybe a quick squirt of Lynx will disguise the smell of stale sweat enough for him to get away with wearing yesterday's shirt for one more day, he is thinking. And he'll put the other two shirts in the laundry basket for Mum to do today, together with his favourite T-shirt that she won't let him wear. It's got JESUS LOVES YOU in big letters on the front and below them, in much smaller letters, BUT EVERYONE ELSE

THINKS YOU'RE A ****. Why is it, he wonders, that his armpits have started to hum so badly?

It's because it wasn't until recently that a particular type of sweat gland called an apocrine gland, liberally distributed around his body wherever there are lots of hair follicles, like his armpits, nipples and private bits, started to react to his 16-year-old boy hormones and add a little oil to his regular sweat. Worse still, he began to produce lots more of this stuff whenever he felt a bit nervous or shy, such as when he was on the receiving end of a good grilling from the school principal or chatting to that blonde girl he really quite fancies. Whenever he feels a little stressed out these days, his armpits start pouring, his forehead becomes dotted with hundreds of little sweat beads and his palms get horribly moist and slippery. He looks and feels like the comedian Lee Evans at the end of one of his brilliant but frenetic stand-up performances. It isn't so bad to begin with. At first he just feels damp and clammy. The water, salt and urea – the same waste product that is filtered out through his kidneys in his pee – don't have any distinctive odour. But as soon as the bacteria which live permanently on the surface of his skin get to work on it, kerr-pow! The by-products of their reaction on his apocrine secretions are truly appalling.

Ben knows that a quick shower would really help, just to rinse off the worst of last night's armpit activity. But he's running short on time. He has still got a bit of that essay to finish off before he hands it in at college today, and Dad's yelling up to him from the kitchen about something and doesn't sound happy at all. Lynx. Where's his Lynx?

Ben immediately assumes Poppy's taken it. He curses again.

In fact, Poppy hasn't nabbed Ben's armpit deodorant at all, which has rolled under his bed and joined the rest of the clutter there. She is oblivious to what her brother is doing because she is much more concerned about the alarming state of her hair.

What is she going to do about it? Mum says that as a foetus Poppy would almost certainly have been covered in fine downy hair called lanugo. Apparently it would have been black, like a sparse layer of fur. Then, about a month before she was born, the lanugo would have been shed into the amniotic fluid in which she was suspended. As a toddler, when Adam and Eve adopted her, her hair was a lovely dark copper colour. Then it gradually got darker.

Here she is now with drab, curly jet-black hair which is totally lifeless, boring and utterly unacceptable. Her hairstyle is going to have to change. How come Ben's hair is thick, dark and uniform? she is thinking. How come Eve's hair is red and glossy and Adam's is, well, receding and greying? Hair is important. Not so much for camouflage, insulation, mate identification and scent scattering, as it is in the animal kingdom. But for most humans it is still arguably their most important natural ornamentation, outwardly and ostentatiously denoting ethnic origin, sex, age, general state of health and personality. That song Poppy had heard in the reprise of the musical *Hair* said it all. How did it go? Something about flamboyant affectations and gaudy plumage, Poppy thinks to herself.

The texture of hair is probably its most amazing

characteristic. Humans can never grow vivid green and purple plumage like birds or exhibit the lurid phosphorescence of exotic fish that dart around in turquoise lagoons. But with the limited ranges of black, brunette, redhead and blond, a terrific impact can still be made with a little bit of creativity and panache. So what determines our natural hair colour, and why does it change as we go through life?

Hair colour is basically dependent on the presence of unique, natural pigments called melanins which exist in the outer part of each hair shaft. The production of the pigments is inherited through at least four different genes and a lot depends on whether you inherit these genes from just one of your parents or both. Most people have black or dark brown hair because the pigment eumelanin is genetically dominant. Lighter and red shades of hair colour are due to the presence of a different kind of melanin called phaeomelanin, which arose thousands of years ago as a mutation of eumelanin. It's common for young children and toddlers to have light or even blond hair, only to see their hair colour darken dramatically during puberty. Nobody quite understands why this happens. Then later in life, as un-pigmented hair becomes more common with advancing years, your hair colour becomes lighter again and greyer, just like Adam's. In fact, it is often said that by the age of 50, 50 per cent of people will have 50 per cent grey hair.

Poppy adores her mother's hair. It's flaming red. Eve told her that the Ancient Greeks thought red hair was a sign of bad luck and even of a barbaric and ruthless nature. These days, people think redheads are just fiery and impetuous.

Lots and lots of fun though, if Eve is anything to go by. Ben's hair is dark. Luckily, Poppy thinks he stands no chance of also becoming tall and handsome, otherwise he'd be insufferable. He'd be swanning around comparing himself to Ben Affleck or Johnny Depp. It's all wasted on him anyway, and he wouldn't have a clue why Poppy might consider it important because he's still a little colour blind, although not as severely as his granddad thanks to his new contact lenses. Poppy on the other hand just wants to be blonde. According to one survey she read recently, two-thirds of British women colour their hair to radically change their appearance. Some just want to emulate their favourite female celebrities, but others, 90 per cent of them in fact, reckon they will succeed better in business, 80 per cent want to boost their self-esteem and confidence and nearly 50 per cent say that they feel sexier for making the change. Other reports suggest that 65 per cent of women go to salons to turn blonde or to highlight, while 68 per cent of TV newsreaders are blonde and 65 per cent of all Miss USAs have been blondes.

Poppy had been interested to discover that pure blonde is actually a very rare natural hair colour, if you take the whole of the world's population into account. How blonde is blonde, though? What are we talking about here? There are lots of shades of blonde: strawberry blonde, ash blonde and bleached blonde. That's the sort Poppy wants to be – bleached blonde – and she is just in the mood to make an appointment at the hairdresser's in the High Street and do something about it.

Chapter 8

AT PRECISELY 7.10AM in the Enniman household, by some strange coincidence, four family members are brushing their teeth. Or in Ruby's case, having them brushed for her by Eve. In different bathrooms throughout the house, a total of 119 teeth of different sizes and in different states of development and health are being cleaned and polished. An odd number, you might think, since it suggests 29.75 teeth each, and 0.75 of a tooth would be horribly jagged and uncomfortable. But Ruby has not yet replaced her primary deciduous teeth or milk teeth with grown-up secondary ones, and Adam has got a full set of wisdom teeth while Ben and Poppy each have at least one un-erupted third molar. Eve's teeth, the one set not being brushed, have always been near perfect, a situation reflecting the loving care she has lavished upon them throughout her life. She will always be grateful for the fact that her uncle Jack was a dentist, and an excellent one at that. From her very first appointment with him, he made Eve's dental check-ups fun.

It was Uncle Jack who originally noticed her first bottom incisor coming through at the front when she was just six months old. It was Uncle Jack who had told Eve's mother how important it was to start brushing that tooth and all the other new ones that would soon be emerging. It was Uncle Jack who had Eve lying back in his dentist's chair at the age of two to get her to look at her own teeth (eleven of them by then) in a mirror so he could teach her the best technique for brushing them. Maybe it was because he had made it so interesting and involved her from the start that visiting the dentist was never something she dreaded or tried to put off to another day. On the other hand, she had never needed a filling or any other invasive treatment. She could only guess that it was these unpleasant experiences that explained dental phobia in others.

Uncle Jack had even given Eve her first paid job in her school holidays, helping out at the dental surgery. She would see all these people coming in holding their aching jaws and looking tortured and terrified in equal measure. People with stained teeth. People with broken and missing teeth. People with no teeth. What could Uncle Jack do for them, she used to wonder. And people with foul breath, brown furry tongues, cracked lips or false teeth connected to pink plastic plates that they would regularly push in and out from the roofs of their mouths with their tongues. She often wondered how her uncle could do his job at all considering how disgusting some of his patients seemed to be. But he had never lost his enthusiasm and had often said that however bad the state of someone's mouth, there was always something he could do to help.

Her own experiences and observations in his surgery certainly reinforced what Uncle Jack had taught her about her own dental hygiene. She must have seen it all. That gradual progressive decay in so many teeth and that terror of the drill. That high-pitched whining noise followed by the low grinding sensation of spinning metal on rotten teeth. Dental caries, tooth abscesses, dry sockets, malocclusion, receding gums, periodontitis, the lot. Tooth decay had been the most common. People were constantly eating loads of sugary foods and drinking lots of sweetened drinks but then not brushing their teeth regularly. Deposits of food, saliva and bacteria, which her uncle told her was called plaque, would build up on the surface of their teeth. The bacteria would work on the sugar to produce an acid which would gradually erode the tooth enamel.

Unsurprisingly, people would come in with toothache that they had endured for weeks for fear of the imagined pain of the treatment. Toothache that people had tried to remedy themselves with massive doses of painkillers and alcohol. Toothache that hadn't responded to cinnamon and oil of cloves. Toothache that had kept people awake all night and given them migraines or driven them to pull out their own tooth with rusty pliers or try that old trick of tying it to a doorknob with string and then slamming the door. Uncle Jack reckoned that in 40 years of dentistry he'd never met anyone for whom that particular little manoeuvre had worked. Yet her uncle had always been so gentle. She never ever saw him hurt a patient in all the time she was with him.

It made her wonder why all these people waited so long to come and get their toothache sorted. It was probably because of everything else, over and above simple toothache, that can affect the teeth and gums, she thought. It wasn't just adults who would come in either; it had been teenagers and small children too. Some of the children were so young they hadn't even got their grown-up teeth yet, but all their milk teeth were brown, soft and discoloured. Their parents had apparently been leaving them in their cots all night with sweetened drinks in feeding bottles to keep them comforted. The constant sugar against their teeth had had disastrous effects, from which many would never fully recover. The hard protective outer covering of the teeth, the enamel, would give way first, creating a small cavity. Left untreated, this would enlarge, leaving the underlying soft dentine in the main bulk of the tooth vulnerable to attack as well. Eventually, the pulp – the living core of the tooth containing all the nerves and blood vessels – would be affected. Exposed to infection, it would ultimately die, leaving dull, lustreless and blackened teeth.

Eve still remembers one particular patient. He was a 27-year-old man with acute necrotizing ulcerative gingivitis (ANUG), otherwise known as trench mouth. His gums were so inflamed that abnormal growth of the bacteria which live in the mouth had occurred and his gums had become red, soft, shiny and very, very swollen. They were covered with a horrible greyish deposit and they bled as soon as Uncle Jack touched them. The stench was appalling and the patient was obviously in a lot of pain. His neck glands were enlarged and he was feverish too. Eve's uncle had sorted him out with a really thorough descale, hydrogen peroxide mouthwash, painkillers and antibiotics, but it was a salutary lesson in the value of looking after one's own teeth and Eve had never missed a toothbrushing since.

The worst patient she ever saw, a lady she has never forgotten, was a rather gruff, smelly, shabbily dressed woman with just two remaining teeth at either side of her lower jaw. How she managed to eat anything solid with them was a complete and utter mystery. These two teeth were in fact the lower canines and whenever she stopped complaining long enough to close her mouth, she would thrust her lower jaw forward so they protruded upwards and came to rest on the outside of her hairy upper lip. She looked every bit like a grumpy warthog and everyone in the surgery was rather scared of her. Sadly she had died at a relatively young age from a heart attack, which Eve remembers Uncle Jack saying wasn't surprising since he was convinced that gum disease and dental caries were inextricably associated with internal illnesses and could shorten someone's life.

Research in the last few years has proven his instincts correct. Inflammation of the gums, which is much more common in smokers, produces toxic substances in the bloodstream that cause further inflammation in the blood vessels around the heart, leading to heart attacks. Living with gum disease, which dentists call gingivitis, carries a staggering 24 per cent increased risk of heart attack, another very good reason for Eve to make sure that Ruby gets her nice little teeth thoroughly cleaned this morning.

'Porridge is good for me, isn't it, Mummy? And it doesn't hurt my teeth?' Ruby asks as her mother stands behind her with a toothbrush at the ready.

'It is, my love, but the sugar isn't good for your teeth.'

'I don't put sugar on it.'

'That's good. Because you don't need it. But there's natural sugar in the banana and in the honey, remember? And that has the same effect on your teeth, which we have to look after.'

'Otherwise they'll fall out?'

'They'll go all soft and discoloured. And they could give you toothache. So best to give them a good brush now.'

The good news is that Ruby has always accepted that brushing her teeth is a necessary ritual. This is helped massively by her latest state-of-the-art rotary toothbrush, which comes with a flashing unit attached to the wall that times how long each part of Ruby's mouth should be brushed for, turning a smiley face into an unhappy one if she brushes too hard and bleeping when she has brushed for long enough. Gadgets are so popular with children these days, Eve thinks, and it seems a shame that so many of them have no educational or practical function. This one does, however. It clearly works for Ruby and is worth the expense for that reason alone. Maybe they'll come up with a gadget soon that can wipe children's noses and make that fun too. She can always live in hope.

As Ruby spits the last traces of toothpaste from her mouth, Eve idly wonders about the baby growing inside her and whether he or she has got any teeth forming yet.

Babies' teeth begin to develop before birth, usually during the fourth month of pregnancy. By the time they are born, all the primary teeth and many of the permanent teeth have started to form, although in most cases they're still invisible, buried deep beneath the gums and out of sight. There are exceptions, though, and funnily enough one of Eve's friends, Janine, had a baby born with a tooth that had already erupted, which made breastfeeding a particularly hazardous experience for her. It was like playing Russian roulette with your nipples, she'd said at the Netmums' group, which had made all the other long-suffering breastfeeding mothers grateful for small mercies.

Eve's unborn baby is deriving all the nutrients for growing its tiny teeth from her own healthy well-balanced diet and, if needed, from the reserves of calcium in her bones. It's good to know, Eve reflects, that all that stuff her uncle taught her when she was young is now paying dividends for her family. She might even have become a dentist or at least a dental nurse herself if a media career hadn't beckoned. She had once been tempted, although in retrospect she is happy with her decision. Which option would most people choose if push came to shove? A TV interview with an A-list actor in a private suite at the Dorchester, or a 30-minute hygienist appointment with a grumpy warthog of a woman in Peckham High Street? Eve's final choice had not been hard to make.

She will never regret the dental knowledge she possesses, though, and she will at least be able to impart her wisdom to her children. Ruby's teeth are going to be as magnificent as her own. Ben and Poppy have both got good teeth, although Ben's had some overcrowding issues and Poppy's had some orthodontic work done to correct a bit of an overbite. She's still wearing the corrective brace her orthodontist had her fitted with and Eve is relieved that she doesn't feel self-conscious about it.

At that exact moment, one floor above Eve, Poppy is in her bedroom brushing her teeth, thinking how awful her 'train track braces' look. Why on earth did she choose pink ones? They make it look as if she's totally pigged out on candyfloss and hasn't bothered to wipe her teeth clean with her own tongue. The braces seemed a good idea at the time. There was a large gap between her front teeth at the top, whereas her side teeth were crowded together. Her lower front teeth were biting into the roof of her mouth a little too. So her mum insisted on taking her to the dentist and he persuaded her to have pink stainless steel brackets bonded to each tooth with springy arch wires attached to them and elastic bands to pull the teeth back into a better alignment. That was a year ago and the dentist is still telling Poppy that after her next adjustments are made, she'll have to continue to wear the train tracks for another six months. Then she'll be wearing a retainer all the time for another three months, and then just at night for three more months. What a commitment, Poppy is thinking. But worth it? she asks herself. Well . . . yes. Definitely yes, when she thinks about it.

She would never admit it, but it's one of the things she is most grateful to her mum for. When Poppy looks at the growing number of 20- and 30-somethings now having it done, she's glad she is getting the treatment out of the way in her teens, along with many of her friends. It's a pain that she has to take even more trouble to keep her teeth scrupulously clean with the braces on, but she knows that if she doesn't, the plaque will accumulate around the brackets and when they finally come off, the surfaces of her teeth would be marked and stained and look even worse than before her treatment started. Sometimes, especially when she's in a hurry, like today before school, she wishes she could just not bother. But she knows she'd only be sabotaging herself.

She needs her teeth. She can't just let them rot. She has been to her biology lessons. She needs them for chopping her food into small pieces to begin the process of digestion. The bite has to be even, and she needs to masticate. Poppy is delighted that her mum actually encourages her to chew gum. Everyone else's parents hate the idea. Eve says as long as Poppy buys sugar-free gum it's fine. It contains xylitol, which is good for teeth and when she chews it massages the gums at the same time. But Poppy isn't doing all this chewing, brushing, flossing, de-scaling and mouthwash business to preserve the health of her teeth, nor chomping into particularly firm Cox's apples to initiate the process of digestion. As far as she's concerned the over-arching function of her teeth is to help her look really *hot*. A good set of teeth which are straight, even and brilliantly white is de rigueur these days if she wants to be taken seriously and enjoy a decent love life. What self-respecting girl is ever going to enjoy lingering kisses if they don't have a flawless and sparkling set of pearls like Scarlett Johansson, Julia

Roberts or Cheryl Cole? And would she ever be tempted to put her tongue down the back of a boy's throat if they had teeth like Austin Powers? Of course not.

So, she is thinking, all you with famously perfect smiles, eat your hearts out. Poppy Enniman is the new kid on the block, or will be as soon as these braces are off. Everyone will be able to see her coming for miles. Her boyfriends had better have pearls to match too. All she asks in a boyfriend is to have clean, brushed hair and teeth, trimmed fingernails, an acne-free face and absence of body odour, earwax and toe-jam. And cracking good looks. And a fab personality. And a good sense of humour. A bit of money wouldn't go amiss too, but now she is just being picky. At least she won't be fooled again by someone like Mark. She had really fancied him at one point, with his broad shoulders, ambling gait, long dark hair and that broad, tight-lipped smile he always used to give her. Then he'd suddenly ruined it all by smiling with his mouth wide open. What a mistake. It revealed a long row of yellow tombstones. For Poppy, it was game over.

When the time comes, should she need it, she'll have no hesitation whatsoever in bleaching her teeth. Success and popularity will be guaranteed. Jobs will beckon. The man of her dreams will appear as if by magic. All she will require is a single visit to the dentist, who will take a quick impression of her teeth and have a dental tray made up for her to hold the 10 per cent concentration hydrogen peroxide tooth bleach. Every night before bed she'll put a pea-sized blob of the gel into each part of the tray which will then cover the outer surface of each tooth, and every night she'll push the tray on to her teeth and leave it there until the morning. Within a week she'll look fantastic. Everyone will want to walk next to her, give her attention and snog her. Well, Ben's friend Frankie might, with any luck. Not for her those toothpastes making implausible claims about tooth whitening. Not for her those watery applications designed to cover up teeth stains for occasional nights out. She is not going to have any teeth stains. She will still drink tea and coffee and she might even have a drop of red wine and a ciggie when her mum isn't around, but she'll see to it religiously that bleaching on a regular basis will prevent any dental discoloration.

Running her tongue sideways from left to right across the bottom surface of her top teeth, she looks in the mirror and imagines a gleaming band of light emanating from her mouth. Simultaneously, Ben is in his own room on the other side of the corridor, rinsing off the toothpaste and thinking roughly the same thoughts as Poppy about how good it feels to have clean teeth. No morning bad breath and no lingering odours attributable to beer, curry, onions or garlic. Come to think of it, he wonders, what, apart from the food that you eat, causes bad breath? So many people seem to have it. And why is it that the people with the worst breath seem to sense you are moving away from them when they speak to you and take a step or two closer to compensate? They invade your space and spew that foul smelly halitosis right into your face. Why do they do that? How can they be so blissfully unaware of it when you yourself just want to gag?

Bad breath, as it happens, is a major issue and one Ben

feels particularly aware of. It isn't just down to the food people consume. That much he knows. If people feast on curries, salami, onions and smoked foods like kippers all washed down with copious amounts of alcohol, they can expect to have bad breath and a mouth which feels like the bottom of a hamster's cage in the morning. And no amount of tooth brushing is going to solve that, since by the morning the problem is no longer just an oral one.

By the next day, the foods have been digested and broken down into chemicals which are excreted from the lungs, not the mouth. It's for this same reason that the police can detect alcohol on people's breath with a breathalyser after they've been drinking. The device measures the alcohol you've consumed by reading your breath. It's the same with food. Garlic especially. Experiments have shown that you can even rub a clove of raw garlic into the soles of your feet and detect it on your breath several hours later. But why would you want to do that? Ben muses. The world already knows about garlic. It's all the other chemicals that make our breath smell we need to know more about, like hydrogen sulphide, methyl mercaptan and putrescine. Putrescine. How aptly named for something so putrid. What was it George Orwell had written in *The Road to Wigan Pier*? 'You can have an affection for a murderer . . . but you cannot have an affection for a man whose breath stinks.' Ben's thoughts precisely.

Everyone has the same basic problem when they wake up in the morning. All the food residues which are missed when you brush your teeth the night before stagnate in the mouth and allow bacteria to get to work on them, breaking them down into pungent chemicals. This is what creates the unpleasant smell. The same thing happens during the day of course, but when you are awake, the natural movement of your tongue and cheeks continually washes away food debris and dead cells, shed from the lining of your mouth, in your saliva. That's the other problem. When you're asleep saliva production slows right down, allowing your tongue and gums to dry out, so the nasty niffs are not rinsed away. It's even worse if you sleep with your mouth open, as anyone with nasal congestion due to colds or allergies inevitably does. It was the same when Ruby had all those problems with her swollen tonsils and adenoids and had to mouth-breathe all the time. Her morning breath was disgusting then, Ben painfully recalls.

What's worse, a couple of his schoolfriends have halitosis. In fact, Ben doesn't really know if he wants to call Tony and Alex proper mates any more. Halitosis is pretty unforgivable in his opinion. In Tony's case, it's probably just because he smokes. Alex's problem is that he never eats after going to the gym. He's really into weightlifting, which he reckons will help pull the girls, but what he doesn't realize is that he uses up all the glucose stored in his muscles in the form of glycogen and after 30 minutes he's beginning to break down the stored fat in his body. After 60 minutes of pumping iron, those breakdown products of fat metabolism – ketones – are his sole source of energy, and unfortunately they don't just circulate in his bloodstream, they're excreted in his breath too. It's a horrible sickly sweet acetone-like smell which you just can't escape if you're within a few feet of him. Like a stink-bomb in the school assembly hall, a little goes a long way. It's called ketosis and any sportsman who needs an energy boost will experience it. As will anyone in their immediate vicinity. Even that rather sweet redhead in year 11 sometimes has a few ketones on her breath. Ben would definitely be up for kissing her otherwise, but one of the reasons she's so slim is probably because she's starving herself. She could do with an energy boost, thinks Ben.

For similar reasons, people who are diabetic develop ketosis too if they're short on insulin. The sugar in their blood can't get into their muscles without insulin, so the muscles and liver break down stored fat into ketones for energy instead. As conscious as Ben is of his own and other people's breath, he feels he needs to know if the toothpaste he has just used has sufficiently addressed the problem.

He knows it's almost impossible for someone with halitosis to be aware of it themselves, as the smell receptors in their noses have become used to it. It's just the same if you get into a really smelly car or toilet. The smell seems appalling at first, but after a while you grow accustomed to it. How else could cheese factory workers, lavatory attendants, pathologists doing post-mortems and zoo-keepers ever do their job properly? So Ben performs that trick he picked up from the Fresh Breath Centre's website. He pokes out his tongue as far as it will go and licks the underside of his left wrist. It doesn't have to be his left wrist, it's just that it's easier for him because he is left-handed. He waits five seconds and then sniffs the place he licked. Hmm. He is not sure. He does it again on the other side. Seems OK.

Great. Clearly he hasn't got any of the dozen or so common causes of permanent halitosis. Gum disease. Dental caries. Nasal congestion. Chronic sinus infection. Enlarged adenoids. Acid reflux. Furred tongue.

Even certain medications may contribute to bad breath, such as antidepressants, which can reduce saliva production and dry out the mouth. And isosorbide dinitrate, which Granddad takes for his angina. But, to Ben's mind, all old people smell a bit anyway. The thing is, he does seem to have a particularly acute sense of smell compared to everyone else, even though it's women who are meant to have more sensitive noses. But everyone's always commenting on how Ben has a habit of sniffing things. He always sniffs his food before he eats anything. And his clean laundry. And the flowers that Mum places around the house. And Nicola's hair when she sits at the desk in front of him at college. He'd better be careful no one notices him doing that, he is thinking. Especially Alex, as apparently he's now going out with her. And he's stacked. How can Nicola not notice his breath? Ben wonders. Maybe she's got anosmia and can't detect the ketones? Maybe she likes them. Maybe she's got ketones on her breath too because she's also on a low-carb diet. That must be it, thinks Ben. It's a coming together of ketones. They're a ketone couple in a convivial conjugation of ketones. Ketone consorts. Soon to be pronounced man and whiff. Ben's stream of consciousness is running wild. Stop, he tells himself. If Alex ever found out what he was thinking he'd be dead meat.

Finally, his personal fresh breath preparations are nearly

done. He has brushed his teeth, flossed in between and scraped the fur off his tongue. All that's left now is to gargle with mouthwash. He mustn't forget the tonsils and oropharynx after all. But which mouthwash? He has accumulated a bigger selection of oral rinses in his bathroom cabinet than Dad has single malts in his drinks cabinet. There's the two-phase Dentyl pH containing those three anti-bacterial agents, natural essential oils, triclosan and cetylpyridinium. The label says these lift, absorb and remove bacteria, food debris and dead skin cells with an effect lasting 18 hours. Then there's Listerine, claiming to be good at reducing gum disease and containing alcohol, but not in sufficient amounts to make the impending car journey to college any more pleasurable. He's got Corsodyl, containing chlorhexidine gluconate, which is meant to be the best anti-bacterial of all. And here in pride of place is RetarDEX with chlorine dioxide, which is designed to get rid of those nasty sulphur chemicals which can poison anybody's breath. Finally, he selects the Dentyl pH. He loves how the two different colours in the bottle merge into one homogenous green as the oil blends with the watery stuff when he shakes it up. He fills the cap to the brim, drinks, throws back his head and gargles. One minute later he spits out. There. Tangy and crisp and fresh as a mountain stream. Pity about his armpits.

Chapter 9

SHAVED, SHOWERED AND dressed for the day, with a few beads of sweat on his brow and (to his mind anyway) his receding but distinguished salt-and-pepper hair neatly coiffured, Adam trudges gingerly downstairs to the kitchen. His back is a bit stiff, his knees creak and his pulled left hamstring muscle still hasn't healed from that protracted game of five-a-side football with his team from work. That was three weeks ago. Why is it that it takes so much longer to recover from exercise as you get older? Why does the stiffness in his muscles take two days to come out when it used to come out the very next day when he was younger? What's all this creaking in his knees about and is he heading for the same sort of hip replacement surgery Eve's mother, Grace, has had at the age of 84? He hopes not. He expects all this aching and stiffness to ease off gradually before he leaves home. His hangover isn't helping, though, and it seems to be getting worse.

In the kitchen, having just yelled at Ben to remove his

muddy trainers from the table, he has popped some slices of bread in the toaster and brewed up cups of tea for himself and Eve. The margarine is there on the table, ready to spread on his toast, but he is having a devil of a job getting the lid off the new jar of marmalade. Lovely stuff, if only he could access it. He has tried to turn the lid with each hand. He has applied a damp cloth for extra grip, and short of locating an industrial monkey wrench it really doesn't seem like his first choice of breakfast is going to happen. But look at what's on the TV. There's a shot of some idiot free-climbing a vertical rock face somewhere in the Lake District, and he's hanging on to this tiny craggy outcrop with just the tips of two fingers of his left hand. The rest of him is dangling below it, suspended 1,000 feet off the ground. Adam bets *he* could prise the lid off this wretched jar of marmalade! How does he do that? What kind of strength does he possess in those two magic digits and has he always had that power or did he develop it along with his mountaineering experience?

The human hand is undoubtedly one of the main evolutionary features which set us humans apart from other animals. That's principally because we have opposable thumbs. We can bring our thumb across the palm of our hand to touch our fingertips. This gives us pincer grip and the ability to pick things up, manipulate them and use tools. But we are not the only primate who can do this – chimpanzees and monkeys can also oppose their thumbs to their index digits. What makes *us* so special is that unlike apes we can also rotate our small and ring fingers across the palm to meet our thumbs. This is because we have a unique flexibility in the metatarsophalangeal joints of the fingers – found in the middle of the palm where the long bones in the hand articulate with the wrist. This is called ulnar opposition and endows us with unparalleled grip, grasp and torque capability. The sort of torque capability you need to open marmalade jars, or not, in Adam's case.

The human hand is a miracle of evolution. Think of all the things that we use it for in a typical day. Look what we can do with it. Turn off the alarm on our mobile phone. Flick on a light switch. Open the bathroom door. Grasp our razor and our toothbrush. Aim what is necessary at the toilet. Massage shampoo into our hair. Dry off with a towel. Do up shirt buttons, fasten belts, tie shoelaces and ties. Open the bread wrapper and margarine tub. Push individual keys on the computer keyboard to check our messages. Open the car door with the remote key and grasp the wheel. Flick a piece of fluff off our lapel. Prise our eyelids open and delicately remove a stray eyelash from our eye. Hold a knife and fork. Pick up a coin off the floor. Pick our nose. And, for some people, hang from an outcrop of craggy rock.

The human hand contains 29 major and minor bones (although as an anatomical aberration some people may have a few more), 29 major joints and at least 123 ligaments. There are 34 muscles which move the finger and thumb, 17 of them in the palm of the hand and 18 in the forearm. There are 48 nerves, with 3 major ones, 24 sensory branches – the ones that feel – and 21 motor branches – the ones that control movement. On top of that, there are 30 arteries, almost as many smaller branches, a large number of veins, which dance around on the back of the hand, and a very interesting pattern of skin creases in the palm, which many people believe can tell us what our future holds.

Anatomically, the hand is a major work of art and all of its structures are controlled by nine individual muscles, which are stimulated by three major hand nerves, which in turn are remotely controlled by your brain. Compared to other parts of your body, the area in your brain focused on your hands is disproportionately large. Nearly an entire quarter of your cerebral cortex (the bit of your brain that controls conscious movement and interprets sensation) is devoted to the control of the muscles of your hands.

Adam certainly had to hand it to the hand. It's got an awful lot going for it. Including language. Think about it. You can hand something to someone when you place it in their care for safe-keeping. A hand can be a special talent, a hired help, a member of a ship's crew, one side of an issue, debate, object or thing, a rotating pointer on the face of a clock, a round of applause, a 4-inch measurement for horses, any kind of physical assistance, anything written manually or a collection of cards in a poker game. Hands also contain fingerprints, those infinitely individual anatomical signatures that are characteristic of just one person. So individual, in fact, that their presence at a crime scene can condemn a suspect to life imprisonment. So now you must look at your hands in a completely different light.

Maybe that is why palmists, crystal gazers, augurers, and fortune-tellers read so much into them, the palms of the hands especially. There's no hair on them, and the skin of the palm is tough and durable but still sensitive. It is anchored to the metacarpal bones below with a tough, fibrous band of tissue called the fascia. This stops it from sliding around like a glove three sizes too large when you're trying to grip something, or to prise the lid off a particularly obstinate jar of marmalade.

Adam's brother Mike found out what it was like to temporarily lose the function of one hand. He'd been sawing a tree branch when the handsaw slipped out of the groove in the wood and ran across the back of his ungloved thumb at its base before he could react. It wasn't a deep wound, but it severed the extensor pollicis longus tendon, and he immediately found he couldn't straighten the end of that thumb. He could bend it and make a fist, because the flexor tendons on the palm side were untouched. But the wretched terminal digit stayed fixed in a bent position and he physically had to straighten it with his right hand to get his left to release things. The specialist hand surgeon had repaired the tendon but he'd had to make a fairly extensive incision along his forearm in order to find it. The tendon had retreated quite a long way because of the action of the muscle in his forearm that worked it.

You would think that such a tendon would be a thick, fibrous, rope-like structure. Not a bit of it. The tendon is surprisingly flat, like a ribbon, and soft. When surgeons join the two severed ends together, they say it's like stitching toothpaste. Which is why they sometimes have to put in z-shaped stitches to give the repair sufficient strength, then immobilize the thumb in a splint for 8–12 weeks to allow for full healing. Even then, with regular physiotherapy and finger exercises, the repair can still come apart again if people are not careful. Surgeons know that lots of self-employed builders and carpenters who need the money try to go back to work too soon and live to regret it.

Other kinds of hand injuries are very common too. One-third of all acute trauma in A&E departments is related to upper limb injuries and two-thirds of these occur during people's working years. One quarter of all athletic injuries involve the hand and wrist, while children under six who innocently like to explore the world around them are at the greatest risk of having their hand crushed or burned.

Adam's brother Mike's left thumb is better now, however. He can happily hold a pint of ale with either hand, which is more than enough for him. Adam wonders what would happen to that insane rock climber on TV, though, if either of the two slim flexor tendons in the fingers of his left hand were to snap under the pressure.

He is still up there, looking remarkably relaxed and calm as he swings about in the breeze. No doubt he's replaying that well-known mountaineering mantra in his head: 'You can never feel truly alive until you've looked certain death full in the face.' Yeah, right, thinks Adam. He's prepared to believe him but he is not making any immediate plans to put the theory to the test.

On the TV in the corner of Adam's kitchen, the show's presenters are marvelling at the strength and agility of the climber. Experts in biomechanics of the hand would be able to inform them that the force produced by the climber's muscles to enable him to cling to that ledge would have to be at least four times the pressure exerted at the fingertips. This is because the weight of his body isn't suspended through his arms and hands in a straight line. The forearm muscles have to work considerably harder to hold the finger and wrist tendons in flexion, as his hand works like a claw to drag him upwards. Quite a lot of force then. But considerably less, Adam thinks, than is needed to open his marmalade jar.

As it happens, Ben has just ambled into the kitchen, moved his filthy trainers off the table and, without any apparent effort whatsoever, twisted off the marmalade lid. What the hell? Is Adam really becoming that weak and pathetic? No, he is convincing himself. It must just be down to technique. Or, more to the point, because he is left-handed. Being left-handed has a lot to answer for. The lids of marmalade jars, just like every other household and office object on the planet, are designed for right-handed people. Most right-handed people grip the jar with their left hand and twist off the lid anti-clockwise with their right hand. Adam, however, has to grab the jar with his right hand and try to twist the lid anti-clockwise with his left. Try it. Your left hand can't twist that way. The wrist can deviate almost 45 degrees to the radial (thumb) side, but only about 10 degrees to the ulnar (little finger) side. Marvellous, thinks Adam. No wonder he can't get the cursed lid off.

Statistically, 80 per cent of humans are right-handed, leaving about 12 per cent of humans left-handed like Adam. The remainder are ambidextrous and can use both hands. This state of affairs is entirely determined by which of the two hemispheres of the brain is dominant. Right-handers are left hemisphere dominant and left-handers are right hemisphere dominant. Bizarrely, some people, whether left- or right-handers, will naturally find it easier to perform certain tasks with one hand and other tasks with the other hand. Changing handedness to remove marmalade lids would be really handy, wouldn't it? thinks Adam.

There is something slightly odd about being left-handed. The English noun 'sinistral' denotes a left-handed person – a throwback to medieval times when people still regarded anyone out of the ordinary as suspicious and believed that left-handed folk were witches or possessed by the devil. Sixty-five per cent of autistic people are said to be left-handed and, according to a report (in the *San Francisco Chronicle*), there is a higher proportion of southpaws among gays and lesbians.

So why are some people left-handed and others aren't? How can 10 per cent of identical twins have different dominant hands? And why do left-handed people often die earlier than right-handed people? Some scientists hypothesize that very subtle brain damage before or during birth may be responsible for left-handedness, just as it is for attention deficit hyperactivity disorder, dyslexia or the abnormal clumsiness in certain children known as dyspraxia. It's a theory certainly compatible with a higher than usual incidence in twins, as twins often develop sharing the same placenta and competing for womb space, which could have a negative impact on one baby's oxygen supply to the brain.

As for why right-handed people live longer than left-handed people, the answer is surprisingly simple. In a world designed exclusively for right-handed people, left-handers have more accidents. Sometimes these accidents are fatal. One recent article published in *New Scientist* quoted the example of the SA-80 assault rifle. Apparently, if you're left-handed and you fire it from your left shoulder, it ejects the used cartridges at great force from the right hand side of the gun, into your left eye. Adam can't help but wonder, if he was stuck on a desert island and only had jars of marmalade to survive on, whether he would simply starve to death. But Ben has opened the marmalade now anyway, and Adam doesn't require any hand-strengthening physiotherapy or hand surgery just yet. Thinking about that, he realizes that Mike's accident with his handsaw was really a narrow escape. Had it been a chainsaw, he could easily have severed his whole forearm in a split second. It happens. People lose bits of their bodies all the time with those things.

It isn't surgically possible to graft a severed head back on, but you can have a go apparently at surgically re-attaching a hand or arm, especially if it belongs to the same person and provided you restore circulation to the limb within a maximum of six hours. Even six hours is pushing it, unless you have packed the limb in ice in order to suppress the metabolic activity of the cells, which, without any available oxygen, will quickly shrivel and die. Alternatively, it is sometimes possible to replace lost limbs such as hands via transplantation, though this is a fairly recent development.

The first hand transplant was performed in 1964 in Ecuador, when the recipient experienced rejection of the transplanted hand within two weeks and had to have it removed. A more widely publicized transplant procedure was carried out on New Zealander Clint Hallam, after he lost his hand in an accident while in prison.

In 1998 a huge international team of specialist doctors and micro-surgeons gathered in Lyon, France, to prepare for the historic operation. The procedure was carried out beautifully in the following order: bone fixation first, then the tendon repair, then artery, nerve and vein repair. The painstaking procedure, including fiddly and meticulous attention to detail and with all the operating teams taking it in turns to work around the patient in shifts, took several hours – much longer, in fact, than the average heart transplant. Unfortunately, although the transplant was a success, the patient himself wasn't too pleased with the end result. He was happy enough with the surgery, he just couldn't get used to the idea of a transplanted hand – someone else's precious paw. After all, who was to know where it had been? What it had done? Whom it had fondly caressed? Whose nose it had picked? Consequently, he failed to follow the strict post-operative immuno-suppressant drug and physiotherapy regime he had been prescribed, and suffered a host-versus-graft rejection. It's likely the surgical team also felt somewhat rejected. At Hallam's own request, the transplanted hand was removed about two and a half years after it had been grafted on.

Since then, several successful transplants have been

carried out, including some on victims who've lost their hands during bomb disposal work and other hazardous activities. In New Jersey in 1999, one such individual who had lost a hand in a firework accident had prolonged success after his transplant and later the same year had the honour of throwing out the ceremonial first baseball pitch for the Philadelphia Phillies. What a result, Adam thinks as he tucks into his breakfast. There are many friends he could name who would gladly sacrifice one hand if it meant that with the other they could open the bowling in an Ashes cricket match or pot the final black in a world snooker tournament. Anything for such a magic moment of fame.

Chapter 10

WITH HER TWIN strings of snot, Ruby has now left Eve's side and is clutching her fluffy rabbit with one hand and holding on to the banisters with the other. She is making her way downstairs to join her daddy at the breakfast table.

The morning sickness Eve was fighting off earlier has unfortunately become a trifle more insistent. She had similar experiences when she was pregnant with Ben but this time the waves of nausea are stronger and more frequent. They're also defying medical description because she isn't just suffering in the morning, the symptoms are lasting all day. Previously the size of a satsuma, her womb has now grown to the size of a Granny Smith apple. Her breasts are fuller as well as slightly tender and the small raised pimples around her nipples, the so-called Montgomery's tubercles, have suddenly become much more prominent and sensitive again. These visibly prominent bumps secrete a special lubricating oil to make the breasts more supple in preparation for breastfeeding. The baby inside her, which doctors for some reason insist on calling a foetus – although Eve thinks that is a horrible word – together with the placenta, are responsible for producing a number of extra hormones like progesterone, oestrogen and human chorionic gonadotrophin, which are the main reason Eve is feeling so queasy. It's all very well her midwife telling her it's a healthy sign of a properly functioning placenta, Eve tells herself, but then it's not her who is having to eat dry biscuits and toast all day to prevent herself from being sick.

Her baby is still only the size of a kiwi fruit. Its proportions are rapidly changing, however, and the length of its body has nearly doubled in the last three weeks, while recently the growth of its lovely head has started to slow down. She cannot feel them yet, but tiny little movements are taking place inside Eve as her baby's arms and legs become more active. Now, by the 12th week of her pregnancy, he or she is opening their mouth in response to touch, sucking their tiny fingers and swallowing over a pint of amniotic fluid every day. Their mouth and nostrils have begun to form, and their eyes have migrated to roughly the place they should be in the head. If she could only take a peep inside with X-ray vision, she would also be able to tell for sure whether she should be calling them he or she.

Her nausea has become truly tiresome. She knew this would probably be the worst time in her pregnancy for nausea, but she is really hoping it won't continue throughout the nine months, as it can in some poor women. The good news, she reminds herself, is that it almost always disappears by about the 16th week and in anything other than extreme cases it never really harms the baby or warrants any medical intervention. The amount she manages to eat is certainly not going to render her baby nutritionally deficient, that much she knows. But, as Adam keeps reminding her, she has to watch her weight. She knows she is only meant to be gaining between 22 and 28lb during the whole pregnancy, and even then most of it in the last half.

She can smell the aroma of toast wafting upstairs from the kitchen, but that isn't as bad as the pungency of the grilled bacon that Ben had been cooking last night. That was like delivering an electric shock directly to the vomiting centre in her brain. Where's Adam with that cup of tea he promised her a while ago? That would help for sure.

Later on, when she joins the rest of the family downstairs, she will have exactly the same breakfast she has carefully prepared for the last few mornings. Banana, a bit of dried apricot, a ginger biscuit and dry toast. No jam or marmalade because it's too sweet and sticky. Bland food, eaten little and often to neutralize her stomach acid, that stuff she can feel right now at the back of her throat. Sceptically, she puts on the acupressure wristband her friend has given her in the hope it might help the nausea and she reaches for her mobile phone where Dr Hilary Jones's excellent iPhone app on pregnancy will surely give her a few more useful tips to try. And there at last, standing in the doorway, is Adam with her mug of tea.

Chapter 11

HOW CAN A banana, a cup of tea and two slices of toast generate with certain and satisfying predictability the need to go to the toilet? Straight after breakfast, within moments of Adam swallowing that last mouthful of food, there it is. That familiar and fairly pressing 'call to stool'. Primitive. Primeval. Animalistic. The call of nature. Like the call of the wild. He won't be in there long. The loo is not somewhere Adam chooses to while away much of his precious time. Unlike the females in the house, Adam doesn't have a problem with pooing. Far from it. For him, it's an automatic, unconscious daily routine that he goes through seven days a week, 365 days of the year. Painless, comfortable, possibly even vaguely pleasurable, it's a function he enjoys with unstinting regularity, not just once a day necessarily but sometimes twice or even more frequently, if that's what nature dictates.

The gastro-colic reflex, as it is known, is for fortunate people like Adam a powerful biological response to eating,

producing an immediate desire to use the loo. It's an automatic nervous-system reflex which occurs two or three times
a day, usually after meals, when the stomach has previously
been empty. The entry of food into the stomach immediately
promotes strong muscular contractions throughout the
large intestine, which work in an insistent and forceful
downward direction so that all intestinal contents are
pushed into the rectum. The resultant distension of the
rectal walls stimulates the sensory nerve receptors located
there to send SOS signals to the brain, which immediately
recognizes the situation and triggers the urgent need to defe-
cate. It's a clear case of mind over faecal matter. But in
certain people, when messages from a distended rectum are
not answered and acted upon, the muscular contractions of
the bowel go into reverse (retrograde peristalsis) and un-
excreted waste products are returned to the colon to
accumulate. Adam has heard about these people and he
thinks it's bizarre. Constipated, tight-arsed and anally
retentive is what he calls them. Little does he know that both
Eve and Poppy are among them.

To Adam, the digestive system is a mystery, a constant
source of fascination and fun. How is it that whatever he
eats is always the same old brown colour when it comes out
at the other end? And what is poo made of, and why does it
reek? he wonders. How does his body extract all the
goodness from the food he is eating, leaving only the waste
he doesn't need and can't use? What causes diarrhoea
and those funny little gurgling noises in his belly (which
doctors call borborygmi, he's discovered), and why does he

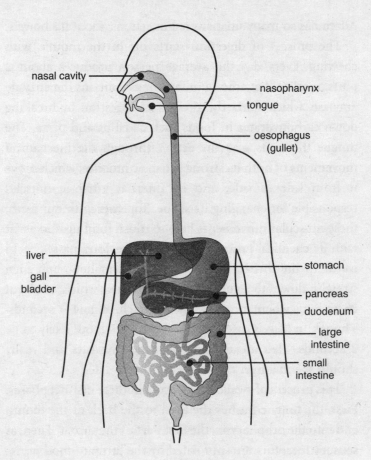

nasal cavity

nasopharynx

tongue

oesophagus
(gullet)

liver

stomach

gall
bladder

pancreas

duodenum

large
intestine

small
intestine

The Digestive System

fart? How much gas does the average person produce
in a day? However much it is, his brother Mike must
produce at least double. It might be something to do with
the beer he drinks, of course. And why do farts smell worse
sometimes and at other times they don't smell at all?

Adam has so many unanswered questions about his bowels.

The process of digestion starts off in the mouth, with chewing. Every day the average person produces about 2 pints of saliva that among other things contains the enzyme amylase, which starts the process of digestion by breaking down carbohydrates in foods such as chips and pizza. The tongue then gets into the action through the mechanical movements of both its strong extrinsic muscles, which move it from side to side, and its internal intrinsic muscles responsible for changing its shape. Together with our teeth, these muscular movements help to mush food up and assist with its chemical processing. From there, food passes backwards to the throat, which is about 12cm long, and then straight down the gullet, or oesophagus, which is about 25cm long, where it reaches the stomach, all in 4–7 seconds. Though in Mike's case, Adam reckons, it's more likely to be 2 seconds. He's an absolute pig when he eats and really should spend longer chewing.

This process of swallowing occurs in three distinct phases. First, the tongue pushes the food to the back of the mouth and into the oropharynx, the first part of the throat. Then, as sensory receptors are stimulated by the lump of food pressing against them, an involuntary swallowing reflex kicks in. At this point, breathing ceases for a brief moment, to allow a mobile flap of tissue called the epiglottis to close off the opening to the windpipe until the food passes further down the gullet. This is to stop the inhalation of food, so people don't choke to death on their burgers. It's a neat little arrangement, and usually works well until the process of swallowing and breathing occurs simultaneously, allowing a large piece of quiche to lodge in the airway. This makes the victim turn blue and requires the Heimlich manoeuvre to clear it. This is the one that Mrs Doubtfire performs in the film of that name by grabbing Pierce Brosnan from behind, wrapping her arms around the lower half of his ribcage, and squeezing sharply upwards and inwards to expel the air in his lungs. When she does this, the offending bit of food is propelled halfway across a busy room of restaurant diners. It saves his life though. It's a result. And a good demonstration of the value of first aid training. The third phase of swallowing is the oesophageal phase. This is where powerful muscular movements in the wall of the gullet (peristalsis) forcibly push the food down towards the stomach. That all happens in a second or two, but the peristaltic movements also occur throughout the gastrointestinal system, all the way from this part of the gut down to the anal canal. Sometimes, however, they work spectacularly efficiently in reverse, causing projectile vomiting. Infants are especially good at this. They can hit the other side of the room 12 feet away if they are given too much milk to drink, too quickly. They also think it's hilarious, which it is unless you're the one clearing up the mess.

The stomach, the next station along the route, is J-shaped, and much higher up than the belly button, where many people imagine it is. In fact it sits immediately below your ribs on the left-hand side rather than bang in the middle. It makes about 2 to 3 litres of gastric juice every day, turning your food into a semi-liquid mushy soup and killing off most bacteria within it with strong hydrochloric acid at a pH of about two or three. This is easily enough to strip paint off pine doors, or remove years of grime from an old coin. Usually, your stomach will hang on to your food for anything between 2 and 5 hours, largely depending on what exactly you've eaten and what you're doing. As with all parts of the digestive system, if you are calm and relaxed it works very efficiently all on its own. If you're busy, stressed out and physically active, however, an involuntary part of the nervous system (the sympathetic nervous system) kicks in and it all but shuts down and stops working. That's exactly why irritable bowel syndrome is worse in people who can never properly relax.

Next stop is the small intestine, which is actually quite big. Or, to be more precise, quite long. It measures about 20 feet from stomach to colon, with an internal diameter of about 2.5cm. All this has to be crammed into the space of the abdominal cavity, so it loops and winds around itself with nautical precision. In this part of the digestive system food is further broken down into its smaller component parts and anything useful is absorbed through millions of tiny finger-like projections called villi. These serve to vastly increase the surface area of the lining of the small intestine, so that all the nutrients your body needs can be absorbed and assimilated with minimal waste.

The large intestine, or colon, is actually, somewhat ironically, smaller than the small intestine. Despite being three times wider, it is only about 5 feet long. Whatever food is left that hasn't been absorbed by the small intestine finds its way here, where up to 6 litres of water can be reabsorbed each day. This is a really important function of the colon and central to water retention and prevention of dehydration in the body. It's also why people with constipation need to drink more water if they want to pass anything other than dry, hard pellet-like droppings. Contrary to popular belief, however, the colon is not just a repository for poo. It is now known for sure that the colon and the billions of friendly bacteria within it are essential for good health and immunity.

A whopping 60 per cent of the immune function of the body emanates from the colon. One hundred trillion bacteria live inside this part of the gut, without which survival would be impossible. Within days of being born, a baby's intestine is colonized with these friendly micro-organisms, the vast majority coming from its mother. All the evidence shows that if you are born through natural means at full term, are breastfed and not given too many anti-biotics, you're much more likely to thrive. Fewer infections and inflammatory bowel conditions, less irritable bowel syndrome, less constipation and possibly even less bowel cancer too. On the other hand if you're born prematurely, by Caesarean section, are bottle-fed and given antibiotics, the gastrointestinal dice are loaded against you from the start.

The gut microbiota is the collection of micro-organisms that inhabit our gastrointestinal tract and is made up of nearly 1,500 species, many of which have not yet been identified. When the characteristics of our gut microflora change – a condition known as dysbiosis – undesirable pathogenic micro-organisms can take over, leading to allergies, overgrowth of yeast, antibiotic-associated diarrhoea and inflammatory bowel conditions such as ulcerative colitis and Crohn's disease. Dysbiosis can occur through either medication, an unnatural diet, excessive hygiene, stress, immunological compromise or environmental factors such as central heating and air conditioning. Consequences can include infectious diarrhoea, necrotizing enterocolitis (which can be fatal in a fifth of babies of very low birth weight), lactose intolerance, gastritis, stomach ulcers and colorectal cancer.

Huge amounts of research are currently being undertaken on probiotics, which are live microbial food supplements, which benefit the human gut by improving the balance of intestinal microbes. Then there are prebiotics. These are the oligosaccharide-based foods that feed the probiotics which already live in the gut and stimulate them to multiply. If that sounds confusing, it's easier to think of the probiotics as the roses already growing in the flowerbed, of which probiotic supplements increase the numbers, and prebiotics as the fertilizer that feeds the roses. In clinical medicine, probiotics and prebiotics are making a big difference already. By keeping pathogenic bacteria at bay, they can prevent and to some extent treat virulent organisms such as Clostridium difficile, norovirus (the winter vomiting bug), and nasty strains of E. coli and salmonella, all of which wreak havoc in elderly patients. Probiotics also produce vitamin K and some of the vitamin B complexes which humans need for life but cannot manufacture themselves.

So there Adam is now. Sitting on his 'throne' dutifully obeying the messages from the sensors in his rectum and the primeval call of the gastro-colic reflex. Is he an observer or a non-observer? That is, does he visually inspect his poo after passing it or not?

Half the nation does, the other half does not. The former tend to live longer because your poo can tell you a surprising amount about your state of health and can sometimes even be instrumental in saving your life. Poo that is pale or clay-coloured can point to malabsorption of fat such as in coeliac disease. Poo streaked with red indicates bleeding from the colon or back passage. Poo as black as tar suggests the presence of digested blood coming from higher up the digestive tract, perhaps from a peptic ulcer or even cancer in the stomach. Motions that float suggest a problem with the digestion of fat. Foul-smelling stools in infants are a feature of cystic fibrosis. Frothy silver-streaked stools are a typical feature of pancreatic cancer, ironically described by doctors as 'a flash in the pan'.

The consistency of poo is pertinent too. Specialists even have a dedicated rating scale for it with seven different rankings. It's called the Bristol Stool Form Scale. Type 1 refers to separate hard little lumps like nuts. Type 2 are sausage-like but knobbly stools. Type 3 is a sausage with cracks in the surface. Type 4 is . . . probably enough! Suffice to say that 7 is totally watery in both consistency and colour: the sort of result you can reasonably expect after a particularly nasty episode of Montezuma's revenge following an injudicious buffet meal at a dodgy roadside café in provincial Egypt.

Adam, as it turns out, is a *casual* observer. For him, and others like him, possessed of a digestive system with lightning-fast bowel transit time, it all happens so rapidly. For him it's a question of sit, drop, stand, wipe, turn, glance, flush. Just like that. Whatever it looks like, though, what does it actually consist of? What are the crucial components of poo?

First, there is any water that hasn't been absorbed by the colon, along with cellulose derived from insoluble dietary fibres. Then there are dead intestinal cells which your bowel lining has shed naturally and a large quantity of bacteria which themselves form a quarter of all the waste you produce. If most people eat on average about 3lb of food each day, or over 1,000lb every year, by the age of 70 it's estimated that they will have passed a grand total of around 33 tons of poo.

So why is it brown? And why does it stink? It's brown because your liver secretes a brown liquid called bile into your gut, which is designed to help with the digestion of fat. The brown colour comes from the breakdown of exhausted red blood cells and the oxygen-carrying pigment they contain called haemoglobin. The smell comes from the bacteria (which are generally pongy wherever you find them in the body) and two particularly stinky chemicals known as skatole and indole. Anyway, for some reason, Adam is grateful for the fact that his poo doesn't seem to smell at all today. Pondering this thought, he goes to the sink, turns on the tap and begins to wash his hands.

Just at that moment, Ruby barges in.

'Ooh, it stinks in here, Daddy,' she says with all the innocent honesty of a typical five-year-old.

'You can talk,' Adam answers back with a grin. 'I'm surprised you can smell anything with all that green stuff coming out of your nose.'

Chapter 12

As ADAM CARRIES little Ruby back into the kitchen to prepare her favourite breakfast of porridge, banana and honey, Ben is sitting at the table, looking at his mobile phone and laughing his head off.

'What's so funny then?' Adam asks.

'Nothing,' Ben shoots back.

'What's making you laugh then?'

'It's just a joke, Dad. It says have you ever gone on to the internet and visited www.conjunctivitis.com?'

'And . . . ?'

'It's a site for sore eyes . . . get it? I thought it was funny.'

What exactly is it that makes us laugh? Adam thinks to himself as he looks blankly at his son. What *is* a laugh? And what possible purpose can it have? Do animals laugh or is it just something we humans do? In physiological terms, a laugh is a series of spasmodic, partially involuntary exhalations of breath with strange vocalizations thrown in, usually signifying amusement. To some extent, it is based on fear and anxiety. Fear, for example, of our own social embarrassment, or somebody else's, fear of loss of dignity, fear of being shown up, mocked, exploited or injured. There's nothing funnier than seeing someone tumble over on a dance floor or step on a tennis ball they haven't seen and collapse in contortions. The Germans even have a name for it. It's Schadenfreude.

This fear of embarrassment and injury is probably why sex and death crop up in so many jokes. And that's just it. Hilarity and harm are very closely linked. It's that old pleasure and pain thing. A great deal of humour arises from the juxtaposition of apparently unrelated objects or ideas, and the element of surprise that creates. Why shouldn't you marry a dwarf with a low IQ? Because it's not big and it's not clever. That kind of thing. The fact that it is a politically incorrect joke just reinforces the hilarity and harm theory. It's the same neuro-psychiatric process that underpins creativity. Does it have a function though? It may be pleasurable to us, at least most of the time, but is it actually good for us? If so, what effect does it have on our body and mind?

When we laugh, electrical impulses are fired off by nerves in our brain. These create chemical chain reactions elsewhere, both in our nervous system and in the rest of our body. Our hormonal system, for example, produces natural tranquillizers and painkillers like dopamine and other endorphins. Digestion is stimulated and promoted. Our pulse and blood pressure drop. Muscular tension is released. We are mentally more stimulated, and so on. And all this happens from the tender age of around 12 weeks, when our mummy or daddy comes right up close, pulls a funny face, and then makes that peculiar whinnying noise with their lips and tickles us.

Many doctors believe that laughing keeps us healthier. There is compelling evidence that laughter genuinely *is* the best medicine. Some hospitals have even adopted 'laughter clinics' to cheer up adult patients with terminal illness and children stuck for weeks on paediatric wards. Research has shown that laughing can also boost immunity by suppressing levels of cortisol (a potent stress hormone) and stimulating clever immune system boosters like immunoglobulin-A and interleukin-2, which can combat cancer and fight upper respiratory tract infections. Optimists, who always look on the bright side of life and see the funny side of everything, generally live longer than pessimists and enjoy a healthier, more fun-filled life. Psychologists have also found a convincing link between laughter and productivity and creativity at work.

Brain-imaging studies have clearly shown that humour stimulates the brain's reward circuit and boosts levels of circulating 'happy hormones' like serotonin. This is the same chemical that pharmaceutical companies use in the antidepressant Prozac and which naturally stimulates motivation and the anticipation of pleasure. Wire up volunteers in experiments with brainwave equipment, show them a funny cartoon or film and ping, ping, ping, a particular part of their brain called the amygdala, central to emotion and feelings of pleasure, lights up like Piccadilly Circus. Laughing might even be useful for people who are sad or who suffer from clinical depression.

For years doctors thought that laughter was totally incompatible with low mood, but recently they have discovered that 'facial feedback', as they call it, means that adopting a forced but happy expression actually *can* lift someone's mood from within. The use of the many different facial muscles involved with smiling and laughter physiologically feeds back to the brain and switches on the neuro-transmitter chemicals that stop us being sad. In the same way you could apparently ask a happy person to adopt an expression of anger or disgust and after a while they are quite likely to feel angry and disgusted. They are certainly more likely to remember and focus on recent events which have had that effect on them. Most practising doctors judiciously employ a measured sense of humour within their consultations to create a warmer rapport with their patients, foster a feeling of closeness between them, dispel ridiculous hypochondriacal ideas and facilitate emotional catharsis. Many doctors use it to alleviate anxiety and tension. Adam's own GP does it all the time. Only the doctor's sense of humour is an acquired taste, and he leaves Adam feeling twice as tense as he was before.

At that moment, as Adam becomes conscious of Ruby staring at him, he cannot resist the temptation to tickle her. Straight away, she collapses in a spasm of giggles and contortions, imploring him to stop but apparently loving it at the same time. Why can tickling be something we both love and hate simultaneously? Most people love being tickled although the stimulus can also be unbearable. The same impulse, if stronger or more protracted, very soon becomes unpleasant. This can result in an increased heart rate, a rise in blood pressure and the triggering of fight or flight reactions. Humans cannot tickle themselves. If there is expectation, it simply doesn't work, just as a knee-jerk reflex doesn't work if you try to perform it on yourself. Tickling usually only works if there is a hint of anxiety and fear about it. Maybe that's where the phrase 'tickled to death' comes from.

So Adam has to be a pretty expert tickler to get his five-year-old to really enjoy it. There has to be a little bit of tension and danger, otherwise she won't laugh. But if he does it too hard for too long, then she will get upset. She's just like everybody else. When he tickles her she will only giggle and laugh if she feels a little bit of apprehension but without feeling threatened. He is trying it now.

'Stop it, Daddy,' says Ruby as she puts one hand on either side of his face and turns his head strongly towards her. 'I want my breakfast.'

Chapter 13

Up in the sanctuary of her bedroom, Poppy's thoughts are distracted from her hair by a more pressing issue. She feels like she needs to go to the loo but she can't. She hasn't actually been for about a week now, and she is feeling rather uncomfortable. She has always had this problem, just like Eve. She doesn't want to keep taking laxatives, which don't really work and end up making her feel even more uncomfortable. But she doesn't want to feel bloated and blocked up like this either. Why can't her bowels behave more like Dad's, she's asking herself. He goes as regular as clockwork at exactly the same time every morning. She can set her watch by his purposeful strides to the loo, where within a matter of seconds he's done what a man has to do and he's back out again with a rather smug expression on his face. Ben's just the same, only smellier.

So why is it, Poppy is wondering, that only women seem to have this problem with constipation? It's horrible. Feeling full, feeling the urge to go. Sitting there patiently waiting. It isn't as if she doesn't eat normally, she eats the same food as everyone else. So where does it all go? What happens in Dad's guts and Ben's guts that doesn't happen in hers? And what can she do to change it?

Why does nobody ever talk about constipation? It's as if it is the last great taboo. Britain seems divided into two camps: those who talk about their bowels the whole time, telling anyone who will listen when they last went or what caused their traveller's diarrhoea, and those who remain steadfastly tight-lipped and never talk about their bodily functions at all.

Then there's the gender division. Men seem to derive constant pleasure from regaling each other with stories of farting and pooing and this ongoing toilet humour remains with them for ever. But they're not the ones with the problem. Women seem to suffer from constipation much more than men, yet they rarely want to share their concerns with either their friends or their doctors.

Part of the problem is psychological: lots of people are conditioned from their earliest years to consider going to the loo dirty and revolting, and perhaps even dangerous. For many, they simply want to keep such embarrassing things strictly private. We cannot see or visualize our bowels, but actually our intestinal function, while hidden away and somewhat mysterious, is essential to good health. We can't see our brains and our hearts either, yet people regard these as life-giving and vital. So is the digestive system. We have a neutral attitude towards other organs such as the kidneys, the liver and the bladder and we are happy so long as they just get on with the job, without creating any problems. But our guts seem to cause us feelings of indignity, shame, fear of disease and embarrassment. The trouble is, the more distressed we become about our bowels, the worse they tend to function.

Ironically, our brain, which in almost all other areas of physiological function smoothly regulates the complex and sophisticated workings of the human body, can actually throw a neurological spanner in the works when it comes to our guts. The influence of our mind can be a very negative one. The brain exerts its effects through a network of nerves universally distributed throughout our gut, which together form something known as the enteric nervous system or ENS. Messages coming down from our brain make the ENS more sensitive by lowering its threshold of reactivity, causing it to respond to weak signals instead of stronger ones. The nerve endings in the wall of our gut become 'upregulated' and go into overdrive. The consequent amplification of signals from the gut makes them shoot right up the spinal cord and back to the brain again. This is how little twinges of discomfort from distension or constipation can so easily become mixed up with emotional feelings and stress. It's all to do with neurochemistry.

Poppy has had this explained to her by her doctor. She is already quite an expert on the ENS, yet she cannot help thinking that the knowledge isn't much use in the short term when it would be oh so satisfying to have a good clearout and start the day fresh. It does, however, provide her with ideas for changing the connection between her feelings and her bowel habits so she can make adjustments in the future. She can, for example, stop eating on the run and give herself longer at the table to digest meals. She can answer the call of nature as soon as she is aware of it rather than postponing it if something else seems more pressing.

The other adjustments she needs to make are dietary. She knows she needs to drink up to two litres of water each day but isn't sure why. It is, in fact, to prevent the colon from doing what it does so efficiently – reabsorbing any water the body still needs and thereby making the motions drier, firmer and harder to pass. She also needs to eat more fibre – of both the soluble and insoluble types to bulk up the motions, and attract water back into them from the bloodstream through the walls of the colon. For this reason, at school today, Poppy will choose the salad option for lunch, with extra fibre in mind. Undoubtedly, many of her classmates will be doing exactly the same thing.

Constipation is very common, and doctors agree that it generally does affect women more than men. It tends to be worse just before their period, or during pregnancy, because then the female sex hormones are at their highest levels in the bloodstream and this seems to be an important factor. Severe continuous constipation affects about one in 200 young women just like Poppy. What happens is that when the motions are small and hard, they don't stretch the rectum enough to create a strong signal to her brain that she needs to go. The 'call to stool', as proctologists call it, is weak. Because it's so weak, the temptation is to not go to the loo, which is bad news. When she ignores these weak signals, these hard pellety motions actually go back up into the colon due to retro-peristalsis.

Poppy drinks litres of water every day and exercises fairly regularly too, but the water never seems to compensate for the exercise and she ends up feeling even more dehydrated. Even laxatives don't work for her. She has tried taking extra dietary fibre, the so-called bulking agents: Fybogel, Regulan, Normacol and Celevac. And none of the traditional herbal remedies even touch it. She does, in fact, suffer from what medical experts in the field of constipation now recognize as being 'laxative-resistant constipation', which has a lot to do with how her emotions and attitudes towards going to the toilet influence the function of her intestine. Once again, it's a case of mind over matter.

It appears that many women can't go to the loo because they're unable to relax the muscles in their pelvic floor which allow the anal canal to open. Their straining when they try to go certainly increases abdominal pressure, but that pressure isn't translated into pushing anything down into the pelvic area. Psychologists explain this as a form of denial or rejection of this part of their anatomy and attribute it to dark thought processes in their unconscious minds. But these are the very people in whom laxatives are ineffective and for whom extra fibre just makes things worse. So what is the solution?

Poppy remembers her doctor talking about a new treatment involving retraining your abdominal and pelvic muscles to get them to cooperate and stop inhibiting each other. But Poppy was less than convinced. No one's going to play around with her pelvic floor, she is thinking, and for the time being she will have to endure another of those gelatinous torpedo-shaped Bisacodyl suppositories that she is currently pushing up her bottom. It won't work very quickly, she knows, and it's likely to give her some difficulty with lower abdominal cramps a little later in the day. Nor will it make her want to eat breakfast, but she knows it makes sense to eat *something* in the hope of stimulating some natural bowel activity, so she washes her hands, ties her hair up, pulls on a jersey and gets ready to join the others downstairs.

Chapter 14

EVE'S MORNING SICKNESS seems to be abating slightly. The cup of tea certainly helped, as did the banana and the two ginger biscuits. Ruby is sitting up nicely at the kitchen table tucking into her usual bowl of porridge, banana and honey with enthusiasm. Eve feels a flush of satisfaction as she realizes that her younger daughter is enjoying a really healthy breakfast. She has long been a big fan of bananas and she always makes sure there are a couple of bunches in the house.

An apple a day keeps the doctor away, right? Forget it. Compared to an apple, a banana has four times the protein, twice the carbohydrate, three times the phosphorus, five times the vitamin A and iron and twice as much of most other vitamins. Not only that, but a banana for breakfast has proven very helpful for Eve's morning sickness by lining her stomach, reducing her stomach acid and helping her sustain normal blood sugar levels for longer. Bananas in fact contain three natural sugars, sucrose, glucose and fructose, all combined with fibre. This provides an instant boost of energy followed by a more sustainable one. It's why many of the world's top athletes favour bananas in training, as just two can provide enough energy for a vigorous hour and a half's workout. Then there's the mood-enhancing effect of bananas. Tryptophan, a type of protein that our body converts into the happy hormone serotonin, is contained within its flesh. Together with the vitamin B6 it also contains, this can be great for premenstrual syndrome or general bad moods.

Adam is also a fan of bananas but for a different reason. He would confirm that bananas are pretty great for curing hangovers. Stick a banana in a blender, add milk, cream and honey, whisk it all up and then swallow with two ibuprofen. This is his sure-fire answer to the sort of pounding headache and nausea he is suffering from right now. The banana raises depleted blood sugar levels and provides potassium for energy. It also lines the stomach, as does the cream. The honey just makes it lovelier so that the sufferer is less likely to bring it up again as they might with other, less palatable hangover cures.

On top of all these virtues, even the banana skin can be of therapeutic use. Many naturopaths recommend treating warts with it. You take a piece of banana skin, place it on the wart with the yellow side out, and fasten it in position overnight with a strip of parcel tape. They also suggest treating the swelling and itching of mosquito bites in the same way. Apparently many people find it surprisingly soothing.

'Why is porridge so good for me, Mummy?' asks Ruby, interrupting Eve's train of thought.

'Because it's got bananas in it, darling.'

'No, not the bananas, the porridge.'

'Ah. Well. Porridge contains oats, which fill you up and help you grow big and strong.'

Ruby, for the moment, seems satisfied with the answer. Right now, she's at that stage where she's incessantly asking questions. Why does she have to brush her teeth? Why does she have to go to school? Why is her nose running? Some of the questions are easy to answer. Some, however, are more difficult for Eve to explain and others are frankly impossible. Why isn't there mouse-flavoured cat food? she asks. Why don't sheep shrink when it rains? Why does the sun make our hair lighter but our skin darker? Where did *that* come from? wonders Eve. How can small children sometimes see life so clearly? And ask such unanswerable questions?

The question about porridge was fairly straightforward, though. Oats are full of fibre, half of which is soluble and the other half is insoluble. The soluble bit is beta-glucans, which lower blood cholesterol. They reduce absorption from your intestine and stop cholesterol being made in the body as well. If you eat one cup of oats every day your cholesterol can come down by about 2 per cent. Cardiologists correlate that to a 4 per cent decrease in risk of coronary artery disease. Porridge also provides energy, reducing the temptation to snack on sweeties or crisps for elevenses, and is good for our stomachs, as the beta-glucan soluble fibre forms a gel which increases the viscosity of the contents of our stomach and small intestine, slowing digestion and prolonging the absorption of carbohydrates, making us feel fuller for longer. The beta-glucan can even boost immunity by helping antibodies navigate to the site of a bacterial infection and eradicate it. Food manufacturers took the stuff out of our diet about 30 years ago, Eve tells Ruby, when everybody started to eat processed food. It comes from bakers' yeast, which the food industry regarded as an unwelcome fungus and largely eradicated from the Western diet. Eve recalls interviewing Dr Paul Clayton – a scientist who believed that this partially explained why, during the same time period, there had been such a massive increase in diseases related to lowered immunity. But it's clear as Eve looks at Ruby that she has totally lost interest in what she's saying.

'And what about my milk? Where does it come from, Mummy?' she asks. But Eve hesitates before giving her an answer. If she tells her it comes from cows it's likely to lead to a whole host of other questions and the answers might just put her off her porridge. So she tells her she gets it from the supermarket and that it's a very good source of protein, zinc and vitamins A, B2 and B12, also providing healthy levels of iodine, niacin and B6. Calcium too, she adds. Lots of calcium, which is very important for Ruby's growing bones and her teeth, Eve tells her, with just one glass a day providing half the amount of calcium a child her age needs to stay healthy. She can get calcium from green leafy vegetables too, she tells her, like broccoli.

'I don't like broccoli,' Ruby says as she wipes her nose with her sleeve.

'I know,' Eve replies, before adding that that is just another benefit of milk, because it means that until people get to enjoy broccoli, they can take all the calcium they need from a different source.

'That's why your nursery used to give you a glass of milk every day right up until your fifth birthday, which nurseries have been doing ever since 1940 during the war. But now the government want to make it harder to do that.'

'Why do they want to make it harder, Mummy?' asks Ruby.

'A very good question,' Eve replies, as her mind starts to wander.

Chapter 15

MEMORY, MEMORY, MEMORY. How is it Adam can remember certain passages from Shakespeare but can't seem to remember where he left his wretched car keys?

'What a piece of work is a man! How noble in reason! how infinite in faculties! in form and moving, how express and admirable! ... The beauty of the world! the paragon of animals!'

Yes, yes, he thinks. He totally agrees with the great poet's sentiments about the wonders of the human body, and shares his sense of amazement and awe. What is so awesome, though, he is thinking, about not being able to locate something as mundane as his own damned car keys? He came home last night and put them down somewhere. They're not on the key rack by the front door. Not in the pocket of his suit. Not by the drinks cupboard. Where the hell are they?

'Darling, have you seen my keys anywhere?' he asks Eve just as she's clearing away the breakfast bowls and telling Ruby to go upstairs and get ready for school.

'Where did you have them last?' she unhelpfully replies.

'Well if I knew that, I wouldn't be asking you where they are now, would I?' What is it about women? he is thinking.

'I'm getting worried about you,' Eve replies. 'You couldn't find your keys yesterday either. Or your passport last week. Or your squash racquet at the weekend. I think your memory is going. Maybe you're becoming senile.'

'That's rich, coming from someone who couldn't even remember to take their contraceptive pill.'

'I did take my contraceptive pill actually. I told you. This is an *immaculate* conception.'

'Yes, it was pretty immaculate. I remember that bit very well.'

'Pity you can't remember where you left your keys though, isn't it?'

Now *that*, Adam says to himself, is precisely why he married Eve. She's witty and she gives as good as she gets. Her memory is also a lot better than her father Joe's, whose deteriorating cognitive function and recall is of increasing concern to both of them. At least he's remembered they're both going to visit him later on this evening. If his memory gets any worse, he'll start forgetting his way home or that he's left the oven on. He could put himself in danger. Then they'll both have to think about long-term elderly care for him.

The thing is, a good memory is something we very much take for granted yet without it we're lost. Impairment of memory is a perplexing and embarrassing disability that can impact severely on so many different aspects of our lives. When we cannot recollect even the most familiar constants in our regular routines, our lives become fractured and we can feel seriously socially handicapped.

Most of us never have to give a second thought to where we left our house keys, what our telephone number is or the names of our work colleagues, friends and relatives. We can state with confidence who the current prime minister is, what day of the week it is and what we were planning to do at the weekend and with whom. But for many, and for a variety of reasons, this simple, previously automatic recall becomes a jumbled and frustrating muddle. For them, it's easy to imagine that this is the inevitable beginning of some terrifying degenerative process that will change their personality, alter their behaviour and ultimately threaten their sanity. But happily, in most cases of simple forgetfulness, there is a benign and innocuous explanation. A temporary lapse in concentration, a minor distraction, a period of stress. In most cases, this faltering memory can be exercised, revived and rehabilitated much like an arthritic joint or an abdominal paunch.

Our memory is a vital and fascinating part of our minds yet there remains so much about it which is medically unexplained and mysterious. We're astounded by stories of people who have extraordinary photographic memories and who can apparently memorize entire telephone directories. We are amazed by those who retain phenomenal memories for particular kinds of information, despite suffering from strokes or conditions such as autism. We are also drawn inexorably to newspaper stories of traumatic amnesia where a person suffering a blow to the head or waking from a coma has lost all recollection of their previous life and identity. It seems we only really appreciate the bewildering power of our memory when something goes wrong and interferes with it. When problems do arise, whether they're minor or major, we know that any of the three main stages of memory can be disrupted. The first-stage registration of information can easily be disturbed when the mind is distracted in someone who is stressed, anxious or depressed. Storage of information, the second stage of memory, can be upset by physical, hormonal or chemical imbalances in the brain. Recall, the third and final stage, can be jeopardized by degenerative processes, inflammation and ageing.

But of all these potential causes of memory loss, perhaps the underlying process we fear most is dementia, of which the most common and familiar type is senile dementia or Alzheimer's disease. Given that up to 30 per cent of people over the age of 85 are affected by this, and since one of the earliest symptoms is that of gradual memory loss, our anxieties about this devastating condition are understandable. But the majority of us will never in fact suffer from dementia. And while a gradual and almost imperceptible loss of memory is a normal part of 'diminishing youth', as Adam's not very funny doctor says, there is much we can do to preserve our retention and recall of information so that we remain bright and alert in all our mental faculties right up until ripe old age. Fortunately, Eve's father's long-term memory is still very good. He can remember his schooldays and his headmaster's name and even the registration plate of his first car. But he can't remember his grandchildren's names, he forgets what shopping he's gone out to buy, and he'll sometimes phone Eve three or four times a day because he's forgotten he's already done it. Adam and Eve worry about her parents, Joe and Grace, but together they will go and see how they're getting on later tonight.

So, why is short-term memory distinct from long-term memory? And why does early dementia affect short-term memory particularly? What actually *is* memory, and what different types are there?

Our memory depends on a whole series of different brain functions but all of them are based on the recreation of past experiences. This is achieved through the synchronous activity of the specific nerve cells that are electrically triggered by the original experience. Having a good memory means we can recognize the people around us, sing along with the words of a song, get ourselves home in the evening or drive a car. Memory is created when nerve cells, called neurons, fire off together in response to a particular experience. When something triggers the same set of neurons to fire off again in a similar way, the original experience is recreated. The recreation can't be *exactly* the same though, otherwise your brain would not be able to distinguish the memory from your current situation. But it is a similar enough chain of events to produce recollection and vivid familiarity.

Short-term memories usually only remain with you for as long as you need them. They're useful, but only for a limited time. Making a mental note of a business colleague's name or phone number is a good example. This is working memory, and the sort that benefits students who cram for their exams the night before and then forget everything they learned within a matter of hours or days. It also in theory enables Adam to remember where he put his keys last night. Long-term memory, on the other hand, is usually laid down more permanently because of associations with strong emotion, deep personal meaning, novelty and constant repetition of recall. There are many different types of memory too. Semantic memory, which recalls straight-forward knowledge such as the fact that the world is round and not flat; procedural memory, which means you can do things like mow the lawn or use a skipping rope; and episodic memory, which relates to previous experiences involving emotion or sensation and which might create a special sense of well-being or excitement when hearing a favourite song, for example. All these different aspects of memory are processed and recalled, not from one central place in your brain, but from many separate regions whose functions are also very disparate. Any particular single memory can, however, involve all these separate areas of the brain simultaneously.

That funny little cat you owned as a child is still clear in your memory but its colour is etched into the colour part of your visual cortex, the strange high-pitched purring it made is stored in the auditory cortex and the acrobatics it performed while jumping from the top of your bedroom wardrobe on to your bed is lodged in the motor area of your brain. All elements of your cat memory are triggered simultaneously and it is this widespread distribution of memories throughout the brain that means they are often still well preserved, even when one aspect of the memory is lost.

It explains why people who incur brain damage following a head injury, stroke or meningitis, for example, will still hang on to certain memories, even though those memories may not be whole and complete. Those people may well be able to remember a face but not a name, a previous residence, but not its location or address.

It was this type of fragmented and partial memory loss that was initially attributed to the notorious 'Piano Man' who was found on a Sussex beach in a drenched dinner jacket, apparently quite unaware of who he was and why he was there. In the days and weeks that followed, the national newspapers were intrigued by the fact that all he was apparently capable of doing was to draw a classical piano, and when one was presented to him, he was able to play with fluency and grace, giving weight to the popular notion that he was a concert pianist who had somehow suffered a sudden and catastrophic loss of memory. As often happens with such newspaper stories, the truth was rather more mundane. Contrary to what was reported, it turned out he was incapable of playing more than a few notes, and after several weeks he finally decided he had had enough of the subterfuge, admitted he was just a regular guy from Germany and discharged himself from the hospital, leaving psychologists and psychiatrists who'd been looking after him somewhat red-faced. His amnesia had been a total fraud. A sham.

Fraudulent pretence of memory loss is not at all uncommon. Ask a guilty child what he or she has done with their sibling's sweeties that they have just eaten and they are likely to reply, 'I can't remember.' It is just as good an excuse for the guilty criminal accused of burglary or murder, who when asked where he was on the night in question replies along the same lines. Not remembering, or at least pretending not to remember, can clearly be exceedingly useful.

But there are many more fascinating cases of genuine amnesia than fraudulent ones, and we are all captivated and intrigued in equal measure when we hear about someone who has recovered from a serious head injury, only to find that they have no recollection of their previous life and identity. They cannot remember their own name, who they are or where they live. They fail to recognize the partner with whom they have shared the same house and bed for 30 or more years. They have to be reintroduced to their own children. The people they work with and the occupation they once had is completely alien to them. This condition, known as retrograde amnesia, means that their memory gap extends backwards from the onset of the original head injury. Yet on another level, they often still know how to walk, to drive and to speak. They can usually also store new information in the period following the head trauma, so that new relationships and new experiences can be learned and retained. Fortunately, in most cases of retrograde amnesia, the problem is mostly a deficit of recall. Somewhere in the deeper recesses of the injured brain where millions of physically shaken-up neurons are dusting themselves down and restoring a sense of order among themselves, the memories that were stored are still there. Gradually, synaptic connections between these nerve cells are remade, dim memories start to become clearer and more reliable again, and that person's memory gap slowly diminishes over time.

Unfortunately, though, none of this helps Adam to find his keys. He can picture them in his mind's eye. Four keys on a big round copper-coloured keyring. One for the car, two for the front door and one for the garage. He can visualize them on the key rack. On the oak chest where he sometimes puts them. On the coffee table. By the kettle. The trouble is, he says to himself, they're not in any of these places. He curses because it's so frustrating. It's such a complete and utter waste of time. He has to get the kids to college and still be at work by nine o'clock. And all the time the traffic will be building up. Calm down, he thinks. He's getting stressed out and it isn't even eight o'clock yet. No doubt Eve will come into the room any second and tell him to chill out and start searching more methodically. Either that or she'll have found them and be waving them triumphantly in front of his nose.

'You won't find them standing around idly like that,' she suddenly blurts out from behind him. 'You'll probably find you have to *move* things. Like lifting up cushions on the sofa or the paperwork on your desk. It's amazing really but some times you'll find things like keys turn up under other things. Try it.'

And there she stands with slightly raised eyebrows and an amused little smile on her lips.

'Why, thank you, my love,' Adam says somewhat sarcastically. 'What would I do without you?'

'Happy hunting,' she answers patronizingly. 'Don't let me distract you.'

And as she leaves, Adam realizes he's just stumbled upon the fundamental cause of his recurrent problem with losing things. Distraction. 'Don't let me distract you,' she had said. And that's just it. He comes home after a busy day at work, his mind totally occupied and his arms encumbered by a briefcase, sports kit, a copy of the local paper and goodness knows what else. He comes through the door, puts stuff down, throws off his coat, the telephone's ringing, Ben's asking for a lift to his mate's house, and Poppy wants to show him her artwork. He's distracted. He doesn't even think about what he is doing with his keys. It never even registers. And if it doesn't register, then obviously he is not going to be able to recall it at a later time. So it's not his memory failing after all. It's his family's fault instead.

Whoever is to blame, it's absolutely true that most of the routine things we do every day leave no permanent trace in our memories. To encode something permanently in our memory or at least etch it there for some worthwhile duration, it really needs to command our full attention or be charged with emotion and meaning in the first place. If the woman of Adam's wildest dreams opened his front door when he came home in the evening and stood there naked while seductively holding out her cupped hands so that he could deposit his car keys, the chances are he would remember where he'd left them. He would remember equally well if he was unlucky enough to receive a massive electric shock because he'd put them down on an exposed live electrical wire. This is because the attention he gives the experience is greater, the emotion that attends it deeper, the sensation more significant and, most importantly of all, because a particular part of his brain called the hippocampus also becomes involved. Neuronal activity in the hippocampus processes these more dramatic experiences and encodes them permanently using a procedure known as long-term potentiation. The data coming in is looped around coils of nerve tissue with each bit of information registering and re-registering not just once but many times over. This is the memory bank's equivalent of winning the jackpot. The most powerful information is played back to those areas of the brain where that information was initially registered. So the sight of that woman of his dreams and her perfume in his nostrils are electrically telegraphed over to his visual and olfactory cortex where they are reawoken as a clear memory of the original experience.

All very interesting, Adam would think. But Carmen Electra wasn't there to take his keys off him last night. More's the pity.

On the other hand, what if he ever wanted to forget something? What a strange concept when most of the time we rack our brains to remember. But we would certainly want to forget traumatic memories. A violent assault. Aggravated burglary. Witnessing a murder. Suffering a miscarriage or delivering a stillborn baby. Those awful memories linger on, as well as bittersweet memories that you partly want to treasure and partly want to bury deep inside you. Post-traumatic stress disorder is not uncommon as a result of all these kinds of experience and those terrifyingly disabling flashbacks can ruin people's lives by preventing them from moving on emotionally. Painful memories are without doubt things you would choose to forget if you could. In extreme cases, individuals have even been known to sub-consciously shut off a traumatic memory, rendering themselves speechless or blind in the process.

Memory is a mysterious and highly complex mental function which scientists still only partially understand. What does seem to be understood, though, is that memory is based on a cascade of messages passed along a given path-way of neurons connected together by synapses (tiny fluid-filled spaces between two nerve cells) where the trans-mission of chemical signals takes place. If this is the case, say some, in theory memories should be transferable by injection from one person to another. And in an attempt to prove it scientists have been trying to achieve just that for years.

Memory transfer experiments actually began as long ago as 1953, at first with work on planarian flatworms. Their tiny brains and nervous systems probably meant they wouldn't have much to remember anyway, but incredibly the experi-ments *did* seem to suggest that the flatworms could learn at least a little from memory transfer. If this really is the case and memories are just made up of physical chemical com-ponents that can be injected from one life form to another, the implications are huge. After one single injection from a hypodermic needle Adam might be able to remember everything that Eve has ever experienced. And vice versa. Or that Nelson Mandela has ever experienced. Or Hugh Hefner. And maybe Ben could forgo the hours needed to commit passages of *Hamlet* to memory for his college studies and simply go for the alternative of an injection. Unfortunately, memory transfer has not yet been perfected. But the research continues.

In fact scientists are nowhere near, so for now at least Adam's going to have to widen the search for his keys. If only he had a photographic memory like the celebrated autistic savant Stephen Wiltshire. He and a few others like him are able to memorize things in fantastic detail and faithfully reproduce them even many years later. After only the briefest of tours of central London, he produced a geographically accurate drawing of Westminster and the river Thames and thousands of the buildings lining its banks in perspective and in perfect detail. Adam would put money on *him* being able to find his car keys in the morning, and performing that essential prerequisite to a lasting marriage . . . remembering his wife's birthday and wedding anniversary.

Chapter 16

I N A MINUTE or two, Eve will be seeing Adam and the older children off at the front door, but she can still make the effort to put on her make-up and smell nice first. Adam is just brushing his teeth (with seven amalgam fillings and three porcelain inlays, she reflects smugly).

Eve knows she won't have to wait too long, however, because Adam's idea of tooth-cleaning is a few cursory horizontal scrapes across the gums with a fiercely stiff bristle head lasting all of 20 seconds. A technique guaranteed to cause gum recession and sensitivity and sizeable dental bills in the future.

Having finished brushing Ruby's teeth, Eve's completed the perfect 2-minute dental routine with her own, and is now sitting in front of the mirror on her dressing table selecting a perfume. Isn't it funny, she thinks, that ever since she's been pregnant she's completely gone off some of her favourite perfumes? Suddenly they smell so sickly and musty. Others that she hasn't used for months seem so much fresher and more invigorating. She was half expecting it, as every time she's been pregnant in the past it has had a similarly strange effect on her sense of smell and taste. Why is it, she wonders, that the fresh smell of coffee she used to love so much every morning now makes her feel instantly queasy? Why has she gone totally off alcohol? Why is the aroma of warm toast perfectly acceptable but the smell of frying bacon reminds her of a freshly butchered animal and makes her physically gag? Why does that all happen just because she's expecting a child and is it anything to do with her morning sickness or the cravings she'll inevitably have for things like Marmite, lemon curd sandwiches, Gentleman's Relish and sherbet fountains?

Her sense of smell depends on chemo-receptors located in the mucous membranes of the lining of her nose. These receptors can respond to a vast number of different molecules but to do so the molecules must first be dissolved in water. This water comes not just from nasal secretions but from saliva as well, which is partly why the senses of smell and taste are so inextricably linked. What we taste is actually about 75 per cent smell, while receptors on different parts of the tongue play a significant role as well.

Not all of the nasal lining can detect smell, however. All of the receptor cells are concentrated in one area measuring about 5 square centimetres high up in the roof of the nasal cavity. Despite that, there are about 10–20 million of them. They are in fact modified nerve cells that are swollen at the lower ends where each one gives off several hair-like projections called cilia, which extend downwards to the surface of the mucous membrane. These cilia contain receptor sites which are stimulated by odorous molecules that waft up the nose, causing them to produce nerve impulses which are transmitted upwards to the brain through tiny perforations in the overlying bone, the cribriform plate, in the base of the skull. Here, just above the bone, are two swollen terminals of the olfactory nerve called the olfactory bulbs. These contain millions of nerve fibres which send any impulses detecting smell upwards to the limbic system in the brain, which is closely associated with memory. It's for this reason that we can store, recognize and remember tens of thousands of different smells. From the limbic system, signals pass to the olfactory region in the cerebral cortex where what you are actually smelling is finally interpreted. From here, other nerve pathways link the olfactory cortex back to the limbic system and then onwards to the hypothalamus, an area of the brain closely associated with hormonal activity, circulatory change and even sexual arousal. This is why your sense of smell can be so evocative of memory, and so central to sexual attraction and libido.

The amazing thing is that all this is going on all the time, around the clock. It is only when a smell irritates us, or we find it particularly pleasurable, that we sit up and take notice. Only when an aroma is interesting or recognizable do we wake up and smell the coffee, so to speak. But how does this happen?

The process of smelling is based on a physical fit between odour molecules and the chemo-receptor sites on the nasal cilia. The receptor sites on some cells will fit, say, with ammonia molecules, while those on other cells will fit with ether molecules. The molecules have to be dissolved in the nasal mucus first before they can stimulate the receptors, but the system's sensitivity is incredible given that stimulation of just four molecules can trigger a familiar smell. In this way, we can distinguish up to 4,000 different odours. Exactly *how* the olfactory nerves in the brain achieve this, however, remains something of a mystery. It isn't just a neat fit between chemical molecules and receptor sites that gives us our sense of smell. Molecules which are odour-producing are generally small but the odour perceived seems to be more dependent on molecular configuration than size. Camphor and hexachlorethane molecules, for example, are physically distinct with different chemical properties yet still smell identical to us humans. Scientists who have tried to categorize our ability to detect smell have separated odours into seven basic kinds – pungent, mint, putrid, musky, floral, ethereal (like dry cleaning fluid) or camphoraceous (like mothballs). They hypothesize that it is various combinations of these molecules that bring about the broad range of odours we are capable of detecting. They may well be right, but whatever the truth, there is no doubting that our sense of smell and the natural smell we ourselves give off, even without perfume, is highly individual.

Ruby's smell is quite unique, and Eve would surely be able to recognize the smell of Ben's or Poppy's bedrooms, even if she was blindfolded. But it's a wonder she can smell anything at all with this nasal congestion that she seems to experience every time she is pregnant.

A woman's sense of smell is thought to be greatest whenever she ovulates, and her nose when she is pregnant is
thought to become hypersensitive, endowing her with an
abnormally good sense of taste, leading to all those bizarre
cravings for things like onions in yoghurt. Or coal. Or old
string. She is supposed to be able to smell fear, happiness and
sexual arousal, although the perception of these emotional
aromas would be subjective and described differently by
different women. According to doctors, this is all due to the
blood vessels in the lining of the nose swelling up under
the influence of higher circulating levels of oestrogen. Eve
might need some convincing.

Her Coco Chanel perfume is nice, but it's barely getting
through her bunged-up nose. She has already done everything she can to minimize her symptoms of nasal congestion
and being the neat and tidy person she is, she folds back the
duvet at the head of the bed to ventilate the mattress and
pillow, and opens the bedroom windows to air the room at
the same time. She was once horrified to hear that there are
2 million house dust mites living in the average person's
mattress, and that the weight of a pillow doubles every six
months as a result of their accumulated droppings.

The house dust mite is indeed ubiquitous. It's present in
everybody's home, regardless of cleanliness and hygiene.
This microscopic eight-legged insect is invisible to the naked
eye, and measures just 0.3 micrometres in size. It feeds off
dead human skin cells and sweat and each dust mite
produces around 20 droppings a day. That's even more than
Adam would produce, Eve is thinking, if he himself was a

house dust mite. Adam's half the problem though. If he didn't sweat so much during the night, maybe some of these dust mites would become dehydrated and die off. But there's no chance of that. She knows it is just the way he's made.

The trouble is, it isn't the dust mite itself to which so many people are allergic. It's their faeces. Or, as most people would prefer to call them, their droppings. The droppings in people's centrally heated, soft-furnished, air-conditioned, cavity-insulated houses quickly become dry and fragmented, crumbling into the atmosphere inside swirling invisible dust clouds, only to be inhaled by unsuspecting victims. Some of the components of the droppings are allergenic to the mucous membranes lining the nose, throat and respiratory airways, leading to perpetual nasal congestion (perennial rhinitis), red itching eyes and, in certain vulnerable people, asthma. This is in fact the most common allergy overall in the UK, and is often confused with hay fever and the nasal congestion associated with almost any other condition including pregnancy.

What should Eve do? Call in the council or pest control? Hopeless. Fumigate the mattress? Unnecessary. Throw away the pillows? Yes. At least every year. Wash all duvet covers and sheets in a 60-degree hot wash every week? Definitely. Buy a hypoallergenic mattress and pillows and place all of Ruby's cuddly toys in the deep freeze every night? Only if her symptoms or anyone else's in the house are significant. Some people Eve knows have gone as far as swapping their carpets for wooden flooring and their curtains for blinds. Some of them have had to take steroids and powerful disease-modifying medicines like methotrexate as well as using their inhalers. Some have attended hospital for desensitizing injections on a fortnightly basis. Clearly, the problems these house dust mites cause are disproportionate to their size. And apart from these mites, it seems that one-quarter of the population is suffering from some form of allergy, with an additional 5 per cent joining the ranks every year.

At least Eve's own symptoms aren't too bad. Over the years she's had plenty of tests. They didn't reveal very much that was helpful, although she did at least learn one important thing: that in randomized clinical medical trials, applied kinesiology, cytotoxic testing, provocation neutralization techniques, vega testing, hair analysis and pulse tests have all failed to demonstrate any medical value whatsoever.

Eve had an NHS skin-prick test about three years ago and the reaction was somewhat disappointing. Just a subtle pink halo around where the doctor scratched her skin and a dismissive shrug of the shoulders from the GP as if to say, 'There's nothing wrong with you. Stop wasting my time.' It wouldn't have been so bad had Eve at least been allergic to something. Cats, dogs, horses. Health care professionals even. The implication had been that she was making a fuss about nothing. They may not have been very sympathetic about her possible allergies, Eve is thinking, but hopefully they *will* be a little more sympathetic about her pregnancy. They tend to be whenever there are two of you involved. Maybe it's just because they're twice as liable in terms of medical negligence if anything goes wrong? she wonders.

But what about her baby? Can he or she detect smell yet?

And if so, what would he or she prefer? When they're born, babies very soon become familiar with the smell of their mother's skin and breast milk, but when does that ability actually develop? When does a baby begin to detect odours? Is it possible at all in the womb, given that they are suspended in a liquid-filled cocoon and surrounded by amniotic fluid?

An unborn baby is floating in fluid rather than air, so it cannot sniff as such, but it can pick up odours that are dissolved within the amniotic fluid. It makes sense given that adults have to dissolve smell molecules in their nasal mucus before the olfactory nerves can become alert to them. If Eve, as the baby's mother, eats spicy or aromatic foods, for instance, the amniotic fluid will alter in chemical composition. It is likely that her own unique but subtle pheromones are also continually present in her amniotic fluid and instrumental in initiating the mother–baby bond after childbirth. Give her baby 24 hours to unclog its sweet little nostrils from all the gunk of childbirth and get some air up them, and he or she is almost certainly beginning to experience the aromas that adults come to know later in life.

In well-documented laboratory experiments, babies' noses have been presented with both lovely fresh odours and horrible rotten ones. They were able to smile when picking up the scent of vanilla, banana, honey and chocolate, while grimacing when exposed to rotten eggs or prawns. They're not so different from us adults really. On top of which, other research has demonstrated that babies can also distinguish between the smell of their own mother and her milk, and the smell of other mothers and their milk. They favour what is recognizable and familiar and since they can so accurately recognize an odour, it means that high-level conscious processing is already occurring inside the baby's brain.

What about taste then? Can Eve's baby taste things as well? We know babies can swallow. When Eve was pregnant with Ben, the ultrasound technician had showed him swallowing the amniotic fluid on the scan. According to some authorities, babies paddle around in a veritable smorgasbord of flavours within their mother's amniotic fluid, including the saltiness of sodium, the sweetness of sucrose and, last but by no means least, the ammonia-like bitterness of their own urine. If Eve as a mother wished to sample her own amniotic fluid, it would in fact taste rather briny, which is perhaps why some babies can be seen pulling faces just after swallowing it on an ultrasound scan. Ruby is five years old now and nothing has changed. She can distinguish tastes all right. And she knows what she likes. And what she *doesn't*. Give her broccoli and she will spit it straight out. How long can a child exist on banana, porridge and honey alone? Eve wonders in despair.

Suddenly, Eve starts to feel all emotional, welling up with tears. Maybe it's because she is worrying about Ruby's selective eating. Or Poppy's calorie counting. Maybe it's just that Adam and the older children are about to leave the house. Why is it that she can suddenly feel so unpredictably and illogically overwhelmed by these concerns? Where does this mother's instinctive desire to wrap everyone up in cotton wool come from? And why do pregnant women react quite so disproportionately to what is going on? It is understandable in people who are under enormous stress. We see it in folk who have suffered brain damage, from either a stroke, Parkinson's disease, head injury or multiple sclerosis. We also see it in children with attention deficit hyperactivity disorder or with other learning disabilities. And you can understand why the 1–2 per cent of women who are unfortunate enough to suffer from severe post-natal depression can become so emotionally up and down, but why in early pregnancy, when Eve is happy to be pregnant and not under obvious stress, are her feelings so out of control?

She is not overly anxious. She is not fearful of childbirth, she has been there, done that and knows she is good at it, and she and Adam have no financial or relationship worries. On top of all that, she has no feelings of inadequacy or worries about being a responsible parent. So in her case, it really must all come down to those wretched hormones, oestrogen and progesterone, which fluctuate so wildly and produce all kinds of strange effects on the neurotransmitters within her brain, which are responsible for her moods. What *are* these hormones and what exactly do they do?

Hormones, in essence, are chemical substances produced in one part of the body and transported by way of the bloodstream to other parts, which are then stimulated into action. In pregnancy, we cannot do without them. They are vital to every single process. Without them, Eve could never have conceived in the first place. Human chorionic gonadotrophin (HCG) formed in the pituitary gland at the base of her brain – a gland described by doctors as 'the conductor of the endochrine orchestra' because of its far-reaching effects – influences her ovaries to secrete more progesterone which in turn stops her periods and relaxes the muscular wall of her womb. Oestrogen prepares and strengthens her uterus for implantation of the fertilized egg, and prepares her breasts for milk production. Oxytocin also stimulates her breast milk glands and triggers those early tightenings of her womb, called Braxton Hicks contractions, which lead up to labour itself. Relaxin softens ligaments in her pelvis and back, and in tissues like those of the neck of the womb, in preparation for the birth. Endorphins – so-called happy hormones – help her cope with stress and pain and give her the feel-good factor experienced in the middle third of her pregnancy.

That's the second trimester, which she'll be entering in about two weeks' time. It can't come soon enough in Eve's opinion. How ridiculous it is that she's crying over a TV ad where a cartoon pussy cat simply looks forlorn, or after watching the *News at Ten* or an appeal for Save the Children. Maybe it's an evolutionary mechanism designed to gear up an expectant mother to protect and nurture her family and the baby she is soon to bring into the world? Maybe it's just a pain in the neck, thinks Eve. As far as she's concerned, that's exactly what it is, and she's now determined to gird her mental loins and pull herself together so that she can go downstairs and make sure the kids have got everything they need for school.

So which perfume might be more suitable than Chanel? She has to make a decision. Apart from identical twins, every individual has a unique smell. A beautifully sweet-smelling rose to one person might be barely discernible to another. We all have blind spots for scents: some people can't smell skunks while others can't smell freesias. A room spray or car deodorizer might smell like pine to one person, but vanilla to somebody else. Neither of them is wrong. It is an issue of interpretation. People often compare the smell of bad breath to acetone or smelly feet to Stilton cheese, and interestingly, the molecules responsible for those smells really are identical to one another. Some people can be acutely aware of their own or somebody else's stale body odour, whereas others simply don't have the chemo-receptors for detecting those smell molecules and are blissfully unaware, even with their faces buried deep within a malodorous armpit. Not a pleasant thought.

Perhaps it is fortunate, then, that we humans keep our brains so busy processing so much other information that our capacity for smell has been progressively diminished. As our cerebral cortices have evolved, the power of our sense of smell has tapered away. And the less we rely on our ability to detect scents, the weaker it becomes. If we don't use it, we lose it. People living in a city, for example, fail to pick up about 70 per cent of ambient aromas, while rural dwellers detect more. Country folk can enjoy the smell of fertilizers and cow dung, of lavender wafting in the breeze and the aroma of freshly cut hay.

Studies of the human genome prove that our smell genes have changed far more rapidly than those of our closest living relatives, the great apes. The sense of smell in humans compared to other species is fairly basic. Us humans sniff less, unless we have a cold, in which case we are sniffing for different reasons. But we *taste* more than other mammals, sending aromas from the back of our throats to our noses as we chew. This ability to savour food as we eat is almost unique to humans, and is known as retro-nasal olfaction. It is the physiological reason why Gordon Ramsay, Jamie Oliver and the Roux brothers make so much money. They cook exquisite food for sure, but they still have to rely on their customers' gustatory and olfactory mechanisms to appreciate it.

The senses of taste and smell are in fact closely related. A malfunction in one can seriously disrupt normal function of the other. We start life, as babies, with about 10,000 taste buds on our tongues and on the sides and roofs of our mouths. This is why infants are so sensitive to and fussy about the tastes and textures of different foods. As we get older, we gradually lose the taste buds that are not located on our tongues and even there the sensitivity diminishes. This partly explains why we favour stronger flavours and spicy foods as time goes by.

Our tongues can detect five basic tastes. Sweet, salt and umami (a meaty, savoury flavour) at the front of the tongue, sour at the sides and bitter at the back. Two distinct, anatomically separate cranial nerves are involved and any abnormality in either will significantly affect our ability to appreciate a varied menu. Other factors impinge on this too. Various medications such as the antibiotic metronidazole can produce a metallic taste in the mouth, while smoking and deficiency in certain vitamins can take away our gustatory abilities as well. Occasionally, the first signs of a brain tumour can manifest as a loss of taste and smell. Fortunately, though, that is rare.

Despite all this, taste is undoubtedly the weakest of our five senses. It is stronger and more sensitive in women than men but still remains the least developed of the various mechanisms whereby we interact with our external environment.

A strong smell, though, is so evocative of memory, especially any memories registered within our brains in the first decade of life. In experiments, humans can recall smells with 65 per cent accuracy after one year, while their visual recall of photographs drops to around 50 per cent after only three months. It's a human characteristic that can be exploited in learning and education. Research has shown that if studying is performed in the presence of an odour, that same odour can enhance the recall of that information at a later date. Teachers have been known to burn chocolate-scented candles in classrooms and then again during children's exams such as SATs in an effort to help them score better marks.

The smell of chocolate might not appeal to Eve right now, but a nice perfume is certainly going to, and will perk her up and leave her feeling clean and fresh – even sexy. It is this connection of fragrance with emotion and mood that perfume manufacturers have commercially exploited so successfully for years. They know that lovely scents make women feel happier. And men are slowly learning that washing away their own nasty smells and splashing on subtle masculine ones may make the women in their lives happier too. The perfume industry makes billions of pounds every year from this fact, and fragrances are an essential requirement for almost every woman on the planet. Yet there are far fewer basic scents than you might imagine.

French perfume databases may include more than 8,000 titles, and modern perfumes can contain around 300 different ingredients, derived from over 6,000 raw materials. Yet it doesn't change the fact that there are still just those seven primary odours – floral, minty, musky, pungent, putrid, ethereal and camphoraceous. What the fragrance industry does with all those ingredients is simply incredible. They have known for years that perfumes can make people feel more confident, attractive and sexy. Tests have proven that certain perfumes can make both men and women perceive a woman as being 12lb lighter and a dress size smaller. The success of perfume manufacturers has even spawned a whole other industry, aromatherapy. Lavender and rosemary essential oils are sold by the gallon to calm and relax people. Peppermint, lemon, eucalyptus, basil, grapefruit and jasmine are all promoted for their energizing effects. Green apple aroma is claimed to control appetite, while others are said to give an impression of a large or a confined space and some scents help send you to sleep or can treat anxiety or depression.

Perfumes give pleasure to millions; it's just the cost of them that doesn't always smell quite right. The most expensive perfume in the world is a case in point. The Clive Christian perfume sells at US$215,000 for a small bottle, with glinting diamonds inlaid in the glass container and a fragrance created from an extract of jasmine, cardamom, bergamot, benzoin, lemon and other citrus fruits. Production was apparently limited to just ten bottles, though, presumably because only ten people could afford it. Les Larmes Sacrées de Thèbes of Baccarat seems a bargain in comparison at just $6,800 and Caron's Poivre even more so at a mere $2,000 for a 2oz bottle. However, the cost and popularity of a perfume does not necessarily rule it out as a potential cause of allergy.

Sensitivities and allergies to many well-known fragrances are relatively common. Chemicals within the perfume such as hydroxynitrilase, eugenol, amyl cinnamic alcohol, benzyl alcohol, cinnamaldehyde, geraniol and natural products such as cloves, cinnamon and nutmeg oil, to name but a few, can all cause reactions in the immune system. Between 2 and 4 per cent of women are thought to be susceptible to a mix of common perfume ingredients and even brands labelled 'fragrance free' can still contain lesser-known fragrancy chemicals that users reading the advertising hype may not recognize. Symptoms can range from runny eyes and nose and headaches right through to tingling of the skin and lips and full-blown anaphylactic shock. Not such a bargain then, if you're forking out a fortune on your favourite smell.

So there Eve sits, hunched over her dressing table as she decides which perfume to wear today. What will be best for an emotional yet stoical mother of three with morning sickness in early pregnancy? Either way, she had better decide soon because it seems like Adam has finally found his car keys and is just about to leave the house with Ben and Poppy.

Chapter 17

ADAM CAN HARDLY believe it. He was searching for his car keys for ages and the wretched things were on the inside of the front door all the time. He *never* leaves them there. He locks the door when he comes in and then hangs the keys up on the key rack with all the others. Only this time he didn't. Maybe he should invest in one of those keyrings that bleeps and whistles to tell you where it is. Only that would probably excite Rufus the dog, who might reasonably assume Adam was about to take him out for his daily walk.

Adam is obviously cross with himself but at least the rest of the family think the whole thing is hilarious and have had a good laugh about age-related memory loss at his expense. Adam himself has worked out by now that mislaying his car keys was due to distraction rather than early dementia. The former means leaving your keys in the front door, the latter means leaving them in the freezer.

For some reason Eve seemed a bit tearful as she kissed him and the children goodbye this morning, he thought, but here the three of them are now, driving up the road on the way to school. The radio is tuned into Adam's favourite station and he's starting to relax.

'What are you doing today then, kids?' he asks.

It's total silence in the back where Ben's sprawled with his eyes closed, and Poppy has done nothing but text on her mobile phone since she got into the passenger seat next to her dad.

'Hello? Is anybody there?'

'Same thing we do every day, Dad,' says Poppy eventually, while Ben yawns indulgently and finally opens his eyes. They're grown-up teenagers now, but still they sulk when it isn't their turn to sit in the front.

'Ben, can you put your seatbelt on please,' Adam gently reminds him. But his son steadfastly ignores him and gazes at the scenery flashing past outside. His eyes dart from side to side as they focus momentarily on each passing tree or vehicle. It's called optokinetic nystagmus, where the eyes repeatedly move slowly in one direction then more rapidly in the other, giving a jerky effect. But there are different kinds of nystagmus and occasionally the cause can be serious. Sometimes people are born with it. Sometimes it develops as a result of some neurological abnormality such as multiple sclerosis, a brain tumour or alcoholism. Sometimes the cause is dizziness due to a disorder of the balance mechanism within the inner ear. Ben's nystagmus, however, is just because he's trying to focus on every passing object outside the car.

'Ben, I said put your seatbelt on. If we have an accident you'll end up in hospital just like Toby did.'

Toby, one of Ben's best mates, had wrapped his car around a tree bordering this road just a few weeks ago, rupturing his spleen and tearing his liver in the process. He'd only just passed his driving test, and his inexperience and slight recklessness had led him to take a corner too fast on a rainy day, and write off his car and nearly himself. He's not long come out of hospital and it was the talking point of the school and the immediate neighbourhood. It happened at a spot about one mile ahead of where they are. And now Adam has raised the topic it finally seems to register with his son, who slowly, rather reluctantly and with an exasperated sigh pulls his seatbelt across his chest and clicks it into place.

'I've got jogger's nipple,' he mumbles. 'It hurts to wear a seatbelt.'

'It's the law,' Adam tells him.

'Not if you've got a medical certificate exempting you from wearing one,' he argues.

'Because of a fractured rib, clavicle or breastbone, maybe,' Adam counters. 'Not for a namby-pamby diagnosis like jogger's nipple.'

'It's a legitimate excuse actually. I looked it up on the internet. It's an inflammation of the nipple caused by repetitive chafing,' Ben reads off his phone screen, 'of the sensitive overlying skin against clothing material such as polyester or nylon. Consequently, it is most often associated with athletes and joggers who suddenly indulge in pro-tracted but unaccustomed exercise without the protective lubrication of Vaseline.'

Wow, thinks Adam. Suddenly his taciturn son has become verbose.

'That would be entirely believable, Ben, but when did you last go jogging?'

'You have to be kidding, Dad,' Ben shoots back indignantly. 'I'm running competitively at school later today and I took Rufus out for a 5-mile run last night. It was freezing cold, raining and no one else wanted to take him out.'

'And what? It was the cold that made your nipples sore? Don't think that would wash with the traffic police as a credible reason not to wear your seatbelt. Besides, we're just coming up to the place where Toby decided to rearrange the anatomy of his car together with most of his abdominal organs. Always put your belt on. Better for your nipple to be a tad sensitive than for your entire body to be totally insensible.'

And that, for a few seconds at least, seems to quieten Ben. The silence is only broken by Poppy, in a philosophical mood.

'Why do men need nipples at all?' she asks. 'I mean Mum needs nipples because she'll probably be breastfeeding again in a few months, but it isn't as if men are ever going to breastfeed, is it? They haven't got boobs.'

'Dad has,' quips Ben.

'Cheers, Ben. For your information, my "moobs" are intensely conditioned, highly toned pectoralis major muscles currently enjoying a temporary resting phase. No need to be envious. If you put your mind to it you too could have a body like mine.'

'Yeah. I would if I sat on my arse all day and drank beer every night.'

Adam sighs. 'Nipples for men,' he explains, 'are merely the vestigial remnants of early development when every embryo begins life as a gender-neutral human being and only develops breasts and the ability to lactate if the influence of oestrogen kicks in and that of the Y-chromosome doesn't.'

More silence. He has definitely caught Ben's attention.

'And just in case you're wondering how I know all this, I didn't look it up on the internet like you would. The nipples and breast tissue, you see, are still present in men, and can always respond to female sex hormones if a man for any reason wants to take them. But usually they are just an evolutionary throwback which, because they contain a modicum of erectile tissue, also serve as an erogenous zone for people who are weird or kinky.'

'That's gross, Dad,' chips in Poppy. 'Can you please stop.' And then, pointing to something ahead of her through the windscreen, 'Look. There's the tree that Toby hit!'

And there, on the other side of the road, is a relatively slender oak tree with tall grasses and bracken flattened and torn from the ground all around it and several distinct indentations in its thick bark. Toby had skidded across the road and into the tree, hitting it more or less head-on at about 65 miles an hour.

Suddenly, the conversation between the three of them becomes animated. But in physiological terms how does a conversation actually take place? Most of us never even think about it. We feel like saying something – we talk to someone, they listen and talk right back. It usually comes naturally and happens automatically. Except, maybe, in Ben's case.

In fact, on the scale of the most complex brain activities any of us ever indulge in, a conversation is right up there at the top. The acts of speaking and listening involve many different areas of the brain, and bring into play many different levels and kinds of cognition. When Adam speaks to Poppy or Ben as his passengers in the car, it takes about 150 milliseconds for the sound of his words to reach their ears, for their ears to react to the stimulus and for the resulting electrical signals to reach their auditory cortices and be registered. The words are then interpreted in a specific part of the brain called Wernicke's area in the left cerebral hemisphere. For full understanding of Adam's words, however, other areas of the brain provide input, especially parts of the right hemisphere concerned with body language, facial expression, speech, tone and rhythm. That's why Adam's using his rear-view driving mirror to look at Ben's reaction to what he is saying to him and the expression on his face when he responds. Is he serious when he claims he has jogger's nipple or just joshing? His body language and tone are vital in helping Adam to interpret what he truly means. This is a skill most of us develop and learn to use. People with autism, who have significant difficulties with social skills and communication, don't seem to have this ability to the same extent. If any of the brain regions necessary for processing spoken words are damaged, a person can be left handicapped by not being able to understand what is being verbally communicated.

The words themselves are one thing, but how are they delivered? Are they delivered in angry or urgent tones? Or is the style calm and informative, matter-of-fact or critical? It is the amygdala which interprets emotional tone and, within 200 milliseconds of the words registering, determines the appropriate emotional response.

Next, the individual words are decoded in Wernicke's area in the left hemisphere, leaving the anterior temporal lobe and inferior frontal cortex in both hemispheres to begin to work out their actual meaning. Then, within 400 to 500 milliseconds, the meaning of the words we are hearing becomes associated with memories in our frontal lobe so that we gain full comprehension. Finally, for conversation to take place, we need to respond and speak back.

The process of speaking begins a quarter of a second before our words are articulated. Our brain first selects the words which will express what it is we want to impart and then we have to convert them into sounds. For most people who are right-handed this will happen in specific language areas in the left side of the brain, but for a minority of people and for people like Ben who are left-handed, they are located on the right side, or spread between both hemispheres. So when Poppy wants Adam to stop going on about nipples being erogenous zones, she first needs to consider words associated with ideas and memories which will act as a cue for the selection of the correct language to express her response. This occurs in the temporal lobe, and happens around 250 milliseconds before she speaks.

Then, in Broca's area, a part of the brain critical to speech and communication, prepared words arrive via a substantial bundle of well-developed nerve fibres known as the arcuate fasciculus. It is this vital pathway, which is so much bigger and more highly evolved in humans than in other species, that provides us with a hugely superior facility for speech and language. The next step is phonology, meaning that the selected words are linked to sounds in Wernicke's area, which is located next to the auditory cortex. A hundred and fifty milliseconds before replying to her father, Poppy's Broca's area will match the sounds of the words she has chosen to use with the specific movements that her tongue, mouth and throat will have to make to articulate them.

Within another 50 milliseconds, those movements are initiated by the motor cortex responsible for the control of those parts of her body. But without the input of the cerebellum at the very back of the brain, any words uttered by Poppy would still just be an indiscernible slur. The cerebellum finely coordinates and precisely directs each of the individual muscle movements into one smooth seamless action. The result is clear, perfect diction. In those of us like Poppy with a dominant left cerebral hemisphere, it is the right cerebellar hemisphere which demonstrates greatest activity during speech, whereas it is the left cerebellar hemisphere which is more active during singing. This may well explain why stammerers can often sing fluently and flawlessly, yet their speech remains halting.

So conversations which we all take so much for granted do in fact require a large number of highly complex activities within many different parts of the brain. Adam may only get a grunt from someone like Ben, especially at this time of the morning, but at least the grunt is an acknowledgement of the fact that he's heard him and is responding in what, for Ben at least, is an appropriate way. Conversations flow better when there is something interesting to talk about, however, and the subject of Toby's torn liver and ruptured spleen is just the thing to capture everyone's attention.

Chapter 18

THE LIVER IS the largest gland in the human body. Positioned to the right-hand side of our body, most of it lies tucked away under the lower ribs, which are there to protect it. Above it is the diaphragm, a tent-like sheet of muscle stretching across the top of the abdomen and separating it from the chest cavity above. Below it are part of the stomach, the gall bladder, bile ducts, duodenum, right colonic flexure of the large intestine, right kidney and adrenal gland. There's a lot going on in this particular part of the anatomical world and the structures are soft and delicate. In particular, the liver is jelly-like and pliable and if you could take it out and squeeze it in your fist it would squish like mashed potato through your fingers. It would also appear deep red in colour because of its very rich blood supply. In fact one fifth of the entire weight of your liver is blood.

Your liver consists of two main lobes separated by ligaments. A large vein carrying blood full of nutrients from your intestine enters the liver on its lower surface and sub-

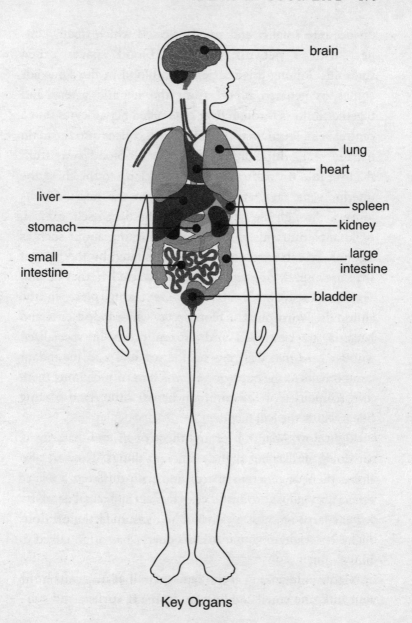

brain

lung

heart

liver

spleen

stomach

kidney

small
intestine

large
intestine

bladder

Key Organs

divides into smaller and smaller vessels which finally connect with a network of small blood spaces called
sinusoids. Joining this nutrient-rich blood in the sinusoids
comes oxygenated blood from the hepatic artery and
together it flows through liver cells called hepatocytes into a
central vein. From there, the blood will collect into the main
hepatic vein, which ultimately drains all blood away from
the liver into the main vein leading back up to the chest, the
inferior vena cava, and finally back to the heart. On its
journey through the hepatocytes in the liver itself, oxygen,
pollutants, nutrients and any toxins lurking about such as
alcohol are extracted and either assimilated by the body if
they are useful, or detoxified and excreted if they're not.
Simultaneously other activities are taking place in the
sinusoids. Worn-out red blood cells, white blood cells and
bacteria are destroyed and broken down in specialized
Kupffer cells that will pass on the residual and redundant
components to the hepatocytes, which in turn channel them
into tributaries of the common hepatic duct transporting
bile towards the gall bladder.

Right now, Adam's liver and those of his two passengers
are doing similar but slightly different things. They are also
doing them at different speeds and with different levels of
efficiency. Adam's, for instance, is busier than usual detoxify-
ing the large amount of alcohol he consumed the previous
night. It's something that Adam's doctor has often talked to
him about.

'Carry on drinking at the rate you're drinking, Adam,' he
had said, 'and you'll end up like George Best.'

What does that mean exactly? Adam had asked himself. Just getting inebriated on a regular basis and behaving very badly? Or needing a liver transplant and total abstinence to survive? Or just plain dead? It was a sobering thought. Literally. It had made Adam wonder about his drinking.

So how exactly is alcohol processed in the liver and why does the liver eventually pack up completely if you keep drinking to excess? When you drink alcohol, it is absorbed in the gut and passes into your bloodstream. On the way, it can cause considerable inflammation to the lining of the gullet and stomach and to the pancreas, but ultimately it travels via a large vein, the portal vein, to the liver. Here, the hepatocytes detoxify the alcohol through an oxidation process and convert it to a chemical called acetaldehyde, which in large amounts makes you feel terrible and is partly responsible for the symptoms of a hangover. Antabuse, a medication often given to alcoholics, interferes with the next metabolic process in the chain and prevents acetaldehyde being oxidized into harmless acetic acid. Antabuse therefore causes a build-up of further acetaldehyde, five or ten times greater than normally occurs if someone continues to drink alcohol. The unpleasant symptoms are designed to deter alcoholics from drinking. Conversely, an enzyme called alcohol dehydrogenase speeds up the detoxification of alcohol within a normal person's liver and any acetaldehyde is converted again to acetyl co-enzyme-A, which the body uses for energy just as it would if it had been produced from the liver, processing carbohydrates, fats and proteins from the food you eat.

The trouble is, there is only a finite amount of alcohol

your liver can deal with, and individuals vary in their capacity to metabolize it. No two people are the same in this regard. In theory, Adam could always tolerate more alcohol than Eve, because his liver is bigger and because males contain proportionately more water and less fat than females so therefore the alcohol is more diluted. But Adam could also be able to tolerate more alcohol than Eve because his regular drinking has led to the production of more of the enzymes in his liver which deal with the breakdown of alcohol. Adam's doctor explained that, for a while, this enzyme induction allows regular heavy drinkers to tolerate a higher intake of alcohol and to withstand its inebriating effects to a greater degree. They therefore sober up more quickly too. But that ability comes at a heavy price.

Excess consumption leads to the formation of collagen and scarring, which forms a barrier between the blood and the liver cells. The hepatocytes cannot function as well as they should and fat builds up within them, leading to both fatty liver syndrome and cirrhosis. That of course is what Adam's one-time hero George Best was diagnosed with – alcoholic cirrhosis of the liver. Now he's dead. As he listened to his doctor's solemn words, Adam had also taken on board that the early changes of cellular scarring were reversible if he drank less. This is because the liver, unlike certain other organs of the body, can regenerate. He's cut down on his drinking, at least during the week, but while he is aware that he ought to try a little harder, he is still drinking considerably more than the 21 units of alcohol per week that the Health Education Authority recommend as the upper limit.

Besides, it is difficult to take his doctor's advice seriously as Adam considers him the biggest consumer of alcohol in the entire village. What was it his doctor had told him in the pub recently? 'The definition of an alcoholic is someone who drinks more than their doctor.' It had made Adam laugh and then counter with that fabulous and time-honoured quote by Madame Jacques Bollinger of the famous Bollinger champagne house: 'I drink my champagne when I'm happy, and when I'm sad. Sometimes I drink it when I'm alone. When I have company, I consider it obligatory. I trifle with it if I'm not hungry and I drink it when I am. Otherwise, I never touch it – unless I'm thirsty.' It is a quote that Adam has applied to his own consumption of not just champagne, but of wine, beer and spirits as well. And so far, to the best of Adam's knowledge, his own liver is not showing any signs of damage.

Adam, like many other people who drink slightly over the limit on a daily basis, is actually in denial about his excessive drinking for a number of reasons. He sees other people doing the same, so it appears to be culturally acceptable behaviour, and he justifies it on the basis that his diet is generally healthy and he exercises regularly. Nevertheless, while he isn't showing any obvious outward signs of alcohol dependence or physical harm other than forgetting where he put his car keys, early scarring and fatty changes in his liver cells are already occurring. A blood sample sent for liver function analysis would show evidence of this. Levels of serum enzymes such as aspartate aminotransferase and gamma glutamyl transferase would be elevated. This would reflect an early degree of hepatocellular injury. Yet other liver function tests including serum bilirubin and protein levels would still be entirely normal, testimony to the fact that the liver is a truly resilient organ and capable of reasonable efficiency, even in extreme cases where three-quarters of its cells have been lost to disease or through surgery.

Of course, liver cells have many functions other than the detoxification of alcohol. Poppy's liver is currently still clearing her blood of the paracetamol she took the previous night for niggling period pain. She had been tempted to take another two within two hours of the first, and more this morning, but little did she realize that a simple overdose of paracetamol such as this can commonly lead to liver damage and even liver failure if excessive intake is repetitive or chronic. As little as 20 normal 500mg tablets of paracetamol in 24 hours has in the past proved fatal as a result of overwhelming and irreversible liver failure. Inside Poppy's liver, proteins are being manufactured to make blood plasma. Albumin, for example, will regulate the fluid balance between blood and the cells of her body. Immune system proteins and coagulation factors which enable the blood to clot are also being produced. Globin, a constituent of haemoglobin, the oxygen-carrying pigment in blood, and cholesterol, which is essential for the transportation of fats around the body, are being manufactured simultaneously. For Ben, part of his breakfast has been converted to glucose but because his body doesn't yet need it for energy, his liver is beginning to store it as a substance called glycogen, which in turn can be converted back to glucose and released back into the bloodstream as and when required.

Healthy livers also regulate the levels of amino acids in the blood, the building blocks of protein for muscles. After a meal, any excess is turned into glucose or proteins while some becomes urea, a waste product filtered by the kidneys and excreted in the urine. The bile produced by the liver is expelled through the common bile duct into the small intestine, where it assists in the emulsification and absorption of fats.

As the liver is one of the busiest organs inside the human body, its diverse functions make it vulnerable to disease and disorders. This is because the liver is the chemical processing factory for everything we eat and every drop of blood in our body. If the food we eat or the blood that passes through it is contaminated, viral inflammation can occur. The viruses causing this inflammation can attack it, leading to hepatitis. Food poisoning, intravenous drug abuse involving dirty needles and even infected blood transfusions can all promote hepatitis in certain circumstances. Bacterial infection can also afflict the liver from underneath if gall stones block the bile duct, leading to inflammation and the spread of germs which normally live within the digestive system. Parasites can literally 'worm' their way into the sinusoids and bile ducts of the liver to cause liver tenderness, engorgement, fever and night sweats, while metabolic disorders can promote the accumulation of minerals such as iron and copper in liver cells, resulting in complications such as liver failure and cancer. The liver cells and bile ducts are also targets for autoimmune conditions where the immune system attacks its own tissues, and because of its rich blood supply the entire organ is a common site too for the spread of malignant tumours from the colon, stomach, breast and bones.

Thankfully, at this moment in time, apart from Adam's liver, which is just about coping with his excessive alcohol intake, the livers of the car's passengers are anatomically perfect. So too is Eve's, who has taken the wise decision to abstain completely from alcohol while she is pregnant. This is partly because she seems to have gone off the taste, and partly because she knows that there is no such thing as an absolutely safe level of alcohol consumption in early pregnancy. The authorities are split on the subject. The World Health Organization says one thing and the Department of Health another, and the Royal College of Obstetricians and Gynaecologists and the Royal College of Nursing have opinions of their own.

It is well known that extremely high levels of alcohol intake lead to foetal alcohol syndrome, the biggest cause of non-genetic mental handicap in the Western world and the only one that is 100 per cent preventable. But even what people call 'normal' everyday drinking can in some instances cause damage. No amount of alcohol has ever been proven to be safe for a developing baby, especially in those earliest embryonic days where essential organs like the liver and brain are rapidly forming.

Babies affected by foetal alcohol syndrome are smaller than average, underweight and have poor muscle tone. Their faces are abnormal, with a typically low nasal bridge, inner eyelid fold, flat mid-face and thin upper lip. The brain may be smaller, mental handicap may be present, the heart and eyes can fail to develop normally, and behavioural problems and impaired social skills are more likely in these children as they grow up. It is a scenario that Eve and many well-informed women like her are all too conscious of, and so she considers her current aversion to alcohol something of a blessing.

When the baby inside her is finally born some six to seven months from now, its liver will be responsible for around 5 per cent of its total body weight. That is why infants generally have such a prominent abdomen. Then gradually as Eve's child grows and their metabolism slows, their liver will become smaller in proportion until in adult life it will represent just 2.5 per cent of its body weight, as Eve's does now. As for Eve's parents, both now in their 80s, their livers have decreased in size and weight as a result of ageing, and because blood flow has decreased concomitantly their livers' ability to handle medication and other drugs has declined as well. Eve doesn't realize it yet, but this is one of the reasons why she feels compelled to visit her parents this very evening, and to try to make sense of her father Joe's disturbing behaviour and confusion. The medicines that he has been prescribed are not being metabolized efficiently by his liver, so they are accumulating in his bloodstream, making him drowsy and forgetful. One small adjustment, such as a lower or less frequent dose, could make a considerable difference.

* * *

Toby used to have a pristine liver. It was busy trying to mop up the few units of alcohol he had consumed at the pub, but it wasn't his liver that was the problem as he approached that fateful bend on the road on the night he had his car accident. It was the effect of the alcohol on his brain that caused the error of judgement.

He was driving on a stretch of road he was familiar with. He'd driven down this very same road about 100 times already if he took into account his driving practice as a learner and his trips out since. But complacency, peer pressure and foolhardiness had combined together and persuaded him to drink. What the hell, he'd thought. He was still celebrating after all. A couple of pints of beer and 'one for the road'. He had underestimated the effect of alcohol on his driving skills. He knew that drinking a lot would impair his reflexes but in his mind he hadn't had a 'lot'. Yes, OK, he'd reasoned, 2 pints was the equivalent of 4 units so the 'one for the road' might have taken the total up to 6 units. But he was young and fit with sharp reflexes and good eyesight, so much better equipped to whizz around the country lanes than some of the wizened old duffers in the neighbourhood who bumbled along like tortoises.

What Toby didn't know was that the particular type of beer he had been drinking was a higher strength alcohol of 5.4 per cent proof, which meant that within the 90-minute period he had been sitting in the pub he'd actually consumed the equivalent of 10 units. His blood alcohol level hadn't even peaked when he hit the tree, as some of it was still being absorbed from his intestine. Had he been breathalysed at that moment, the light would have stayed green and Toby would, at least in a court of law, have been able to argue that he was not driving under the influence of alcohol. Legally, he would have been correct. Medically, however, he would have been totally and utterly wrong.

It is undoubtedly true and very widely accepted that significant consumption of alcohol seriously handicaps a person's ability to drive. Nobody would be happy to be driven by somebody with slurred speech who was unable to walk in a straight line, or who kept falling over and couldn't think clearly. Coordination and balance are visibly sacrificed when blood alcohol levels are high. What is generally less well recognized, yet clearly demonstrated in many well-designed and elegant research experiments, is that impairment of most of the important skills of driving can occur at even very low blood alcohol concentrations. Driving a vehicle involves psycho-motor skills, vision, perception, steering, attention and information processing and all of these functions are impeded by even small quantities of alcohol, albeit to different degrees. Ironically, chronic heavy drinkers can sometimes bluff their way through the traditional field sobriety tests which involve tiptoeing along a white line or balancing on one foot, even at very high concentrations of alcohol. But their vision, perception and finer motor skills are shot right through.

Toby's blood alcohol level at the time of his accident was 75 milligrams of alcohol per 100 millilitres of blood. Legally permissible. But as the level had been steadily rising since he left the pub, his mood had become increasingly disinhibited.

He had felt simply marvellous. He had had a great craic with his mates, and told a few brilliant one-liners. His favourite was about the travelling circus's failure to recruit someone to be their new human cannonball. They couldn't find anyone of the right calibre. He found that hilarious. Soon after, he was off to see his girlfriend, in his own car, with the windows wound down and the stereo blazing. Without him being the least bit aware of it, his vision was changing.

The frequency of his eye movements and the duration of each fixation or 'look' was different. Increasingly, his gaze was fixated in the centre of the driving scene in front of him rather than the periphery. The movement of his eyes and his ability to focus on objects were still relatively unaffected by his drinking, yet his visual field was more tunnelled and similar to that of someone suffering from glaucoma. His eyes themselves were functioning normally but his brain's control of them was slowly going awry.

Toby's brain was also somewhat distracted by a myriad of different sensory inputs all cascading into his head at once: the sound of the car stereo, the wind rushing past the window, the revving of the engine, the street lights strobing between the rows of trees by the roadside and flickering across his eyes. The perceptual complexity of all this arriving at a staccato rate from several sources simultaneously became even more challenging due to the neurologically depressing effect of the alcohol. The tracking of his car as he steered it was imperceptibly worse. He didn't know it, wasn't aware of it in the slightest, but he was becoming more erratic in his lane control and was veering closer to the central white line in the road or to the kerbside as he turned corners. Correction was slower, and over-steer more measurable. Response times to visual stimuli had lengthened considerably. It was taking Toby marginally longer to read each street sign and react to various traffic signals, and as a result, his information processing was clumsy.

It was partly because of this, partly because his reaction times were slower, and partly because he was so focused on his central field of vision that he failed to see the Bambi-like fawn as it sprang out of the bushes and on to the road until several milliseconds later than he would have done had he not been drinking. It was only a matter of milliseconds, but unfortunately it was enough. Suddenly over-steering, as a natural reaction to the presence of the animal and in an impulsive attempt to avoid it, the wheels locked as he over-applied the brakes and he lost all control of the vehicle's path. A shower earlier had made the road wet and slippery, with the rotting leaves of winter compounding the problem. Toby's recollections of everything that happened next were a blur.

The car slewed across the road, mounted the kerb, ploughed through the long grass and bushes and hit the tree more or less head-on. The seatbelt saved his life, but there was no airbag in the ancient second-hand car he was driving, and the force of the impact drove the engine block backwards into the dashboard area. This made the steering wheel thrust hard against Toby's ribcage, breaking several ribs in multiple places. The lower ribs, which are mobile and springy and are only attached to the breastbone by cartilage, were pushed hard back against Toby's backbone, rupturing his spleen and tearing his liver as they travelled. Several of the higher ribs, which are all bone and firmly fixed rather than partly cartilaginous and mobile, fractured, the sharp spicule of bone at their broken ends cutting like a serrated kitchen knife through soft tissue. Toby at this point knew nothing about it. He was concussed and only semi-conscious.

One of the most useful features of the human spleen is that it is not an essential organ at all. If it is surgically removed its various activities can easily be taken over by other parts of the body, such as the lymphatic system and liver. Weighing around 200 grams, the spleen is a spongy fist-sized dark purple organ lying on the opposite side of the abdomen from the liver, behind the lower ribs. It is covered by a fibrous capsule and is filled with red and white blood cells called lymphocytes and phagocytes which eat up foreign particles and bacteria and form an important part of the immune system. It has a rich blood supply coming from a major artery which divides and branches extensively inside it.

So what does it do then if it is not absolutely necessary for survival? The spleen does two main things. First, it regulates the quality of circulating red blood cells by removing and breaking down those that are at the end of their 120-day shelf life. It will also filter out those that are misshapen or defective. Its second role is to help fight infection by manufacturing antibodies and lymphocytes. It is an important but not vital part of the reticuloendothelial system that also incorporates the lymph glands, bone marrow and liver, whose functions overlap.

In the embryo growing inside Eve's uterus, the foetus's spleen is actually producing red blood cells as well, a function that will cease and be assumed by its bone marrow the moment it is born. But Toby's adult spleen was doing exactly what anyone would expect it to be doing at the moment of impact with the tree. It wasn't swollen, as it often can be in teenagers by infections such as glandular fever, nor was it enlarged as a result of leukaemia, lymphoma, sickle cell anaemia or tuberculosis. Any of these would have made Toby's spleen much more vulnerable to blunt trauma. But even a normal spleen can be ruptured by a severe blow to the abdomen such as in a fall, a particularly aggressive rugby tackle or, as in Toby's case, a car accident.

'Toby was so lucky,' says Poppy rather thoughtfully as the three of them drive past the site of the accident.

'Lucky? What's lucky about ploughing your precious first-ever car into a tree and liquidizing the contents of your abdomen?'

'Ben, he could have died.'

'Yeah, but he could also have recovered from his skid, quickly straightened up and continued his journey without any hassle, before meeting up with the winner of this year's Rear of the Year award. That would have been marginally luckier, wouldn't it? Why is it that whenever you hear about anyone falling 12 floors from a high-rise balcony or being savaged by a great white shark and surviving, they are always described as lucky? It's bloody unlucky if you ask me.' To which Adam just laughs out loud.

'He was lucky to survive, I mean,' Poppy adds quietly. 'He lost loads of blood. They took his spleen out. They had to patch up his liver and two perforations in his intestine. And he got septicaemia and nearly died!'

'It could have been worse,' muses Ben.

'How?'

'It could have been *me*,' to which Adam bursts out laughing again and Poppy tries not to smile.

In reality, Toby had been very lucky to survive. Both his ruptured spleen and lacerated liver had haemorrhaged severely. By the time the ambulance had got him to the hospital his blood pressure had dropped to almost undetectable levels. When the surgeons opened his abdomen over 4 pints of O rhesus negative blood was slopping around his internal organs. The worst of the bleeding was coming from the torn spleen so the surgeons attended to that first. Under general anaesthetic a horizontal incision was made across the top of Toby's abdomen, exposing both the liver and the spleen. Speed was essential in a case like this, so a carefully planned, neat little keyhole incision was not an option. The scalpel slashed across the belly in a fashion any Samurai warrior would have been proud of. It was brutal and it was brief. Inside, copious suction of extravasated blood enabled the surgeons to see the operating field they needed to work on more clearly and all major bleeding sites were identified and clamped. Oozing from smaller vessels and the broken surface of the liver itself persisted unabated.

But the priority was to stem any further blood loss from major arteries and then from the 'bleeders', the smaller arterioles which send tiny little spurts of bright red blood into the air. Up to now, as fast as the doctors could transfuse blood into Toby's collapsing arm and leg veins, it was being pumped straight back out again by his injured viscera. It took an hour or so before the surgeons got a handle on it. One after the other, major bleeders were clamped and tied off. Toby's blood pressure increased and his heart rate slowed. The process of repair and reconstruction could begin.

Over the next four hours attachments of other organs to the spleen were divided and the splenic artery itself was clamped and cut. The damaged non-viable parts of the liver were removed and the raw oozing areas coagulated and diathermied with a neat little tissue-ablation device called a CUSA. The two traumatic holes in the tranverse colon and small intestine were explored, cleaned up and sutured. Finally, when the heroics were completed and the consultant surgeons had left the operating theatre to write up the case, the more junior staff closed the wound and got the patient to the recovery room.

Toby had been a fit, healthy young man before the accident and this certainly contributed to his survival and swift recovery. There was a blip when he spiked a temperature and the microbiologist became excited about some coliform bacteria cultured from his blood samples. But these were the bugs that had escaped from his perforated gut into his abdominal cavity and the problem had been anticipated.

Consequently, intravenous antibiotics were soon able to mop them up and eradicate them. In an age when antibiotic resistance is already one of the greatest threats to the general survival of mankind, that was a relief for everyone concerned.

Eighteen days after writing off his car and very nearly his own life, Toby was discharged from hospital and sent home. While the parents of Toby's close friends used the experience to lecture their own children on the dangers of reckless and drunken driving, Toby revelled in his new-found notoriety. He would dine out on the story for years.

Chapter 19

JUST AT THAT moment, Poppy's mobile phone rings. It is actually a message from Ben sent half an hour earlier, screaming at her to bring back his deodorant that she'd never borrowed in the first place. She listens just long enough to hear his unfounded accusations and pick up the high-pitched whining tone of his voice and then decides to play the whole thing on loudspeaker to show her brother and dad just what an idiot Ben can sound like sometimes. She hadn't been anywhere near his smelly room, she wouldn't have touched his dreadful choice of vulgar sweet-smelling boys' anti-perspirant with a bargepole anyway and he could only have been about ten strides from her own room yet still decided to phone her. How lazy was that?

The message, much to Ben's discomfiture, is clumsy and unjustified. But he is more concerned and somewhat disconcerted by the tone of his own voice as it's played back to him. Does he really sound like that? he wonders. He had always assumed he had a deeper, less nasal tone to his speaking voice, with clearer diction. Why is it, he is thinking, that whenever he hears his recorded voice it always sounds so different? He had once asked one of his friends, as they listened to a recording of Ben's acting in the school play, if that was a faithful replication of the voice his friend heard when Ben spoke. 'Absolutely,' had been the answer. 'That sounds exactly like you.' Why is it that all of us sound so different when we hear our own voices played back?

It is all to do with the conduction of sound and the difference between air conduction and bone conduction. When we hear other people speak (or any extraneous sound for that matter), the sound waves travel along our ear canal to our eardrum. This then sends the signals through the bones of the middle ear cavity to the inner ear, from where the messages travel along the acoustic nerve to the brain. When we ourselves speak, however, the vibrations from the vocal cords in our larynx are transmitted to the bones in our skull, including the petrous temporal bone in which our inner ear is seated. These sounds reach there a moment before the sounds coming through the air and drown them out. The result is a sound with which we are much more familiar, the sound we hear every single day of our lives whenever we speak. That strange tinny sound coming from Poppy's mobile is something else completely. It may be the noise other people hear whenever Ben converses with them, but it isn't a noise he recognizes or likes. And coupled with his distinctly uncool message he finds it strangely unnerving. But now Ben's attention is returned to the previous topic of conversation.

'Let what happened to Toby be a salutary lesson to you,'

Adam is saying to Ben and Poppy, who are aware of their dad droning on about things they consider blindingly obvious but are really only half listening. Poppy is more concerned about whether her friend Liz was cross with her over some throwaway remark she made about her diet yesterday and Ben is thinking about the cross-country race he'll be running this afternoon. They don't need to be told that Toby had nearly bled to death as a result of his stupidity. They certainly aren't planning on emulating his achievement any time soon. Over the last few weeks they have both thought long and hard about what had happened. Their mates had worried about whether Toby would survive and reliable information from the hospital had been scarce. Patient confidentiality precluded any news bulletins and the family's privacy obviously had to be respected at that difficult time, so everybody hung on to every bit of gossip and rumour that was going round. Facebook and Twitter had gone nuclear on the subject. Fortunately, Toby hadn't bled to death. But it had been a close-run thing.

How is it that the precious sticky red fluid we call blood, without which we cannot live, circulates so fluently within our arteries and veins, but then spontaneously clots when exposed to the air? Why doesn't it always clot quickly enough to stop some people from haemorrhaging to death, and conversely, why don't unwanted clots form in our blood vessels all the time? The genetically linked blood of Adam, Ben, Eve and Ruby is at this moment circulating around their bodies in a perfect state of equilibrium. To clot or not to clot. It is a question their blood is eminently capable of answering every single day of their lives. Until, that is, external events take over. The solidification of blood, or lack of it, is one of the human body's most incredible features.

Blood starts to clot within a few seconds of us suffering an injury. When a blood vessel is damaged, the clot serves to plug the leak. At the same time, the blood vessel itself constricts to minimize further blood loss. But while blood coagulation in the right circumstances can save our lives, it can also threaten our very existence in the wrong ones. Blood clots, or thrombi, which occur in any major blood vessel of the body, can kill us. A thrombus in a coronary artery will cause a heart attack, a clot in the cerebral circulation a stroke. If it happens in arteries inside the gut or leg, the death of a section of bowel or a gangrenous limb will result. So we don't want our blood to clot too readily, nor do we want to wait too long for it to coagulate when we're bleeding, which is why two basic events need to take place to bring about clotting.

At the site of an injury, tiny fragments in the blood called platelets have to become activated and gather together in clumps. Then chemicals called thromboplastins released by the platelets and any damaged blood vessels stimulate proteins in the blood called clotting factors, which result in the formation of fibrin filaments at the site of the injury. These enmesh with the red and white blood cells to plug any breach in the wall of the blood vessel. But then, if our bodies are knocked around, pressurized and injured all the time, what stops blood from clotting continuously?

To counter the potential for blood to clot inappropriately,

other mechanisms are constantly ready to inhibit it. These anti-clotting mechanisms depend largely on prostacyclin, a substance produced by the walls of healthy blood vessels which inhibits the activation of platelets. On top of that, certain proteins circulating in the bloodstream have the ability to neutralize activated clotting factors. And when the cascade of reactions occurs to bring about clotting, another opposing group of enzymes is activated to produce fibrinolysis or breakdown of the fibrin filaments. Blood flow itself plays a part. By continuously rinsing away any active coagulation factors from places where they have formed, it also reduces the potential of the blood to clot. The discarded coagulation factors are later deactivated by the liver.

In Toby's case, his platelets were well and truly activated. His clotting mechanism did its best. For each molecule of coagulation factor activated by his platelets at the beginning of the sequence of coagulation reactions, 30,000 molecules of fibrin rushed to the site of each of his injuries. There was no shortage of damaged blood vessels to initiate the clotting process. Unfortunately for him, the extent and nature of the injuries meant that his clotting processes were overwhelmed.

Ironically, when larger arteries are traumatically severed, say the femoral artery supplying the whole leg or the radial artery at the wrist, the artery can contract sufficiently to prevent further blood loss. This is due to their thicker, elastic muscular wall. Smaller arteries and veins do not have the same capacity, nor do the capillaries and sinusoids in a contused and lacerated liver. The images depicted in films like *Saving Private Ryan*, where wounded soldiers are seen picking up their traumatically amputated limbs and carrying them to a place of safety, are in fact accurate. The expected torrential blood loss does not always ensue if major arteries are cut through.

But even in normal circumstances and without any internal or external injury, as in the case of Adam, Ben and Poppy right now, there exists within the blood a delicate homeostatic balance of its tendency to either clot or stay fluid. It is called coagulability. As well as traumatic injury, a number of factors can tip the balance either way. Sitting in the car for example, just like travelling on an aeroplane or lying in a hospital bed for protracted periods, makes the circulation sluggish and the blood more likely to clot. This inactivity is one of the principal and most preventable causes of deep vein thrombosis. During her pregnancy Eve is more susceptible to clots as a result of raised oestrogen levels. So too, although to a much lesser extent, is Poppy, who is taking an oral contraceptive pill for her acne.

Eve's father is also more at risk as the common heart rhythm disorder he has developed, atrial fibrillation, has caused a more turbulent blood flow through his heart. He doesn't know it, but his atrial fibrillation is increasing his risk of experiencing a stroke by seven times that of someone without the condition. It also explains why his doctor is treating him with an anticoagulant. The warfarin he is taking – rat poison as he calls it – actually works by interfering with the action of vitamin K, which is necessary for the production of the clotting factors. But because of the delicate balance between the doses he needs to take and its effects – he neither wants too little anticoagulation nor too much – he has to go to the hospital for a blood test every few days to monitor what is happening. It's tiresome and time-consuming for all concerned, especially Eve, who generally takes him, but his doctor has just started talking about a new therapy which does exactly the same thing but apparently doesn't require these regular blood tests. As far as Eve is concerned, that would be a brilliant result all round.

Many other disorders can interfere with the clotting process too. Some people inherit a coagulation defect, such as haemophilia, Christmas disease or von Willebrand disease, all of which cause an increased tendency to bleed too readily and haemorrhage either after injury or even spontaneously. Because their inheritance is sex-linked, the first two of these conditions only affect males. In each of them, just one of the coagulation factor proteins is absent or only present in very small quantities, but without it the whole cascade sequence of coagulation is hampered. In the UK, around 5,000 people have haemophilia, and 1,000 have Christmas disease.

The surgeons who operated on Toby were mightily relieved to discover that he wasn't among them. Had that been the case, there wouldn't have been nearly enough blood within the entire hospital with which to transfuse him. As it was, he received over 30 pints. Other coagulation disorders can be acquired through illness. Severe liver disease or intestinal conditions can prevent the absorption and use of vitamin K. Then there is disseminated intravascular coagulation (or DIC), which Toby's surgeons feared he might be developing while he was still on the operating table. In a case like this where so many thousands of tiny blood vessels have been damaged simultaneously, all the platelets and coagulation factors can be used up and assimilated quicker than they can be replaced by the stricken liver. In that situation, Toby's blood loss would have continued unabated. Fortunately, as the surgeons fought to bring everything under control, the bleeding did eventually settle. Again, Toby's relative youth and level of general fitness stood him in good stead.

Not having yielded to the temptation of smoking, as some of his friends had done, also helped Toby. Toxins in cigarette smoke are well known to make platelets stickier and more likely to clump together. It's one of the main reasons why heart attacks and strokes are much more common in smokers. These illnesses are caused by blood clots in the arteries supplying the heart muscle or brain, and blood clots thrive on sticky platelets. At certain points in Toby's dramatic surgical journey, stickier platelets might have been welcome. At others, they might have been disastrous. In terms of surgical treatment generally, smoking is very bad news. Given the combination of blood vessel injury, post-operative immobility and anti-inflammatory medication, the last thing anyone in hospital needs is increased platelet stickiness, which would make the chances of a deep vein thrombosis even greater.

Suddenly Ben is vaguely aware that his dad has stopped lecturing them about road safety. Adam's car has come to a stop.

Chapter 20

Back at home, Eve is now washed and dressed and ready to do battle with the day ahead. Thank God for make-up, she's thinking. What would women like her do without it? So good for morale and self-esteem. So good for blemishes and wrinkles. It's only a pity it can't do anything to prevent stretch marks, she thinks as she wistfully imagines the size she will be when she approaches the third trimester of her pregnancy. Right now, though, she's got to apply Ruby's eczema cream, dress her, get her ready for school, sort out Rufus the dog, check the venue for the film shoot this afternoon, organize Ruby's pick-up and get herself off to work. First things first though, where *is* Ruby?

There she is, lying on the floor of the lounge downstairs, giving Rufus the Airedale terrier his customary morning cuddle. It's amazing how tolerant Rufus is of the man-handling he receives from Ruby, as she throws her arms around him, squeezes him, pats him on his flat little snout and pulls his ears back. When she and her mum visited the

Blue Cross dogs' home, Ruby had fallen in love with him at first sight.

There he had been, in his pen at the rescue home, standing alert, upright and perky despite having been rejected by his previous owners on the grounds that their child had ostensibly been allergic to him. With his noble oblong head tilted slightly to one side, and his long muzzle and jet-black nose, he stood there with his small dark eyes fixed firmly on Ruby. His mouth was slightly agape, as if he was grinning, and occasionally his teeth would meet in an even bite at the front. His V-shaped ears were folded slightly to the side of his head, and forward towards his happy little face. His chest was deep, with the top line of his back level, and his front legs perfectly straight and covered with a wiry tan-coloured coat. His ears were darker and the hair on his back, sides and the upper part of his body was more of a dark grizzle. Considering Airedales are supposed to be the largest of the terrier family, Rufus was surprisingly small and compact, like a permanent puppy.

Ruby had not been able to talk about anything else for days until they had finally brought Rufus home. Eve had first wanted Ruby to spend some time with the dog to ensure that she would not show any signs of allergy herself, although the Blue Cross had reassured her that in the absence of asthma or a family history of allergy, and with this particular breed of dog in mind, any future problem was unlikely. Eve had pointed out the slight eczema that Ruby has had since the age of 18 months, and which still produces some red scaly areas on her wrists and arms. Instead of Rufus, she could, they had said, opt for a Labradoodle which was even less likely to shed hair and cause problems. But although Rufus's coat was hard, dense and wiry, they said it was not particularly allergenic, and they thought the previous owner's assertions that allergy was the reason for them giving the dog up were almost certainly spurious. Their loss was little Ruby's gain. It proved a match made in heaven and they were immediately inseparable.

Rufus took to Ruby as if they were brother and sister and was unstintingly gentle and protective. Friendly enough with strangers, the Airedale was bright, intelligent, responsive and obedient to the last. He was also tirelessly curious. He would poke his flattened head with its straggly little beard into Eve's cupboards and drawers and was forever looking into nooks and crannies throughout the house. And while Ruby would throw his ball for him to retrieve in the garden and get him to run loyally beside her as she rode her tricycle, she began from the very first day to keep him in line with firm consistent handling, to which Rufus was contentedly responsive. It was a mystery where she had found this instinctive confidence. It was as if it had been inborn. It's a wonderful relationship that clearly provides enormous pleasure on both sides. Although Ruby's eczema is still a little bit of a problem, there is no way on earth she would ever agree to being separated from the dog. But to what extent is Rufus really contributing to her chronic skin condition, if at all? How much animal dander or skin scales is he actually shedding into that lovely shag pile carpet Eve chose, and is the carpet rather than the dog hair causing the problem anyway? How could Eve possibly know?

In the grand scheme of things, Eve reckoned that Rufus's skin scales and hair proteins were not going to have a major impact on Ruby's mild eczema. When you think of all the infections or parasitic diseases that animals can transfer to humans, this was insignificant in comparison. Unwormed dogs can transmit the parasitic infection toxocariasis through their faeces, which can contaminate human hands and every year renders about 100 children in the UK partially or totally blind. Flea bites are itchy and unpleasant too, and cats can transmit toxoplasmosis which Eve is particularly conscious of because this protozoal infection can infect unborn babies and cause miscarriage or stillbirth. In babies who survive, there may be enlargement of the liver and spleen, hydrocephalus, blindness, mental retardation and death in infancy. Cats, lovely though they are, can also transmit cat scratch fever and fungal infections. When you think about it, the list of zoonotic infections, or parasitic diseases that can be transferred to humans from animals, can be quite alarming. Brucellosis can be picked up from cows, goats and pigs, and rats can transmit leptospirosis, otherwise known as blackwater fever. Mosquitoes, when they bite us, can give us malaria, yellow fever and dengue fever, and even harmless-looking parrots can cause psittacosis, a nasty type of pneumonia. Dogs like Rufus, Eve considered, were relatively low risk. Unless they have rabies, which Rufus doesn't.

Most people know that dogs, cats, horses and donkeys are the animals which are most likely to cause allergies like eczema, and she had tried her best to interest Ruby in a parrot, a pair of lovebirds, a tortoise, a hamster or a goldfish. Anything but a dog. Even a decent-sized aquarium with a variety of brightly coloured exotic fish hadn't swayed her. Ruby's heart had always been set on a dog and as soon as she laid eyes on Rufus there was no contest. Eve had plenty of reservations, but ultimately Ruby was insistent. After taking some trouble to assess to what extent Ruby's eczema might have been affected, Eve had concluded that it hadn't made any appreciable difference.

It's the proteins in the fur, saliva and even the urine of a dog to which people with a predisposition to allergy can react. When the animal protein comes into contact with the skin, redness and itching can develop or, if already there, become worse. The itching leads to scratching. That in turn provokes more itching once the immediate but temporary relief from the scratching has passed. It's a phenomenon widely known by dermatologists as the itch–scratch cycle. Ruby could stroke the dog then transfer the dog's proteins to her eyelids, a particularly delicate part of her skin. That could lead to puffy, red and itchy lids. Her skin might become blotchy all over if she started to scratch. Unless the dog was kept outside all the time, its proteins could remain in the house for months if not years, even after a deep clean.

Despite all that, Eve felt a lot more optimistic when she heard that eczema and other allergic problems with dogs were usually worse in the second year of a child's life and not at Ruby's age of five. That was also consistent with what the doctor had told Eve. There was a good chance that in time

Ruby would grow out of her eczema too. Some 50–90 per cent of children apparently grow out of it altogether by their teens and even in the ones who don't, skin inflammation is usually much improved and only flares up occasionally under intense provocation.

Eve realized that the cause of allergies was still little understood. If she believed everything she read about sensitization to allergens occurring within the womb, she would have become totally paranoid by now. Plenty of academic research pointed to the fact that high exposure to indoor allergens could sensitize a growing baby in the womb, even without maternal sensitization. But on the other hand, she had also read scientific articles stating that exposure to animals in early life actually reduced the incidence of allergy as children grew older. Wasn't that why the children of farmers, who were exposed to a huge variety of animal proteins straight after birth, had a much lower incidence of asthma, eczema and hay fever? It all seemed so confusing.

The thing was, Eve had done everything humanly possible to eradicate the dust mite issue already. What on earth was she thinking of by introducing a hair-shedding, skin-licking dog into the home? Ruby of course didn't see it that way and on the whole, with her emollient lotions in the bathwater, her various creams, moisturizers and ointments, Ruby's skin was, if anything, gradually getting better. It really did look as if she was already growing out of her dermatitis. What with Ruby's eczema, Poppy's acne and her own increasingly visible stretch marks, common conditions that affect most families to some extent, it got Eve thinking just how much we take the appearance and function of our skin for granted.

Our skin is after all our largest organ. Weighing around 9lb and containing 11 million blood vessels, it covers an average surface area of about 21 square feet. How else could a homicidal monster like Hannibal Lecter have made lampshades out of his victims' skin, as Ben had pointed out in a recent biology essay.

When not in the hands of the cannibalistic Lecter, however, the skin is a living structure made up of several layers all performing many different functions, some of which are essential for life. The outer layer of cells in the epidermis is dead, with each of us shedding around 40,000 of them every single minute. In fact, most humans shed a whole layer of the epidermis – the outer covering of skin – every 24 hours with the skin in its entirety renewing itself every 28 days. Globally, these dead skin cells amount to about a billion tons of dust blowing about randomly in the atmosphere. But new cells are constantly being produced in the layers of the epidermis to replace the ones that are lost. In the deeper layers of the skin, the dermis, nerve endings, blood vessels and sweat glands are found. From it, on a really hot day, 3 gallons of sweat can be produced to allow the body to cool. The dermis contains fatty tissue as well, which serves to insulate the body as well as act as a shock absorber and energy reserve.

On top of this, our skin also protects us. It keeps what should be kept inside us in, and guards us against dangers from the external environment. As such, it is a reliable, robust and adaptable organ. Some parts of it, like our eyelids, are thin and very delicate, yet even in these places there is still a tough, fibrous protein layer on the surface called keratin, which is also present in our hair and our nails. The barrier our skin provides keeps out all kinds of potential hazards, germs, poisons, toxins, chemicals, insects and radiation to name just a few. None of these is able to reach the bloodstream as a result. When we bathe, our bodies do not fill up with water because our skin is not permeable to it. Yet at the same time we can sweat, and lose waste products and excess heat in the process. The skin keeps our physiological balance under tight control. Our skin also gives us our sophisticated sense of touch. We are responsive to feelings of pressure, heat, vibration and of course pain. Every fraction of a second our skin is sending billions of electrical signals to our brain, where they are registered and interpreted to give us a kind of global sensory picture of where we are and what we are in contact with. It tells us too if we are in any danger. For instance, if Ruby's bathwater is too hot, Eve's hand will soon tell her. As will her lips on the rim of a hot cup of tea or her fingers as she scrapes the ice off her car windscreen in the morning.

The ability to appreciate sensation comes from various receptors in the skin, each individually attuned to detect things such as pressure, temperature or pain. In some parts of the body there is a particularly dense collection of these receptors – in the fingertips, eyelids or genital area, for example – where the skin is acutely sensitive, whereas in other areas, such as the small of the back or the elbow, receptors are sparse and the skin is much less sensitive.

Right now, Poppy is just about to jump out of Adam's car. She will stand talking to her friends outside her school, shivering and shuffling from one foot to another in an attempt to keep warm. She steadfastly refuses to wear what her mum calls a proper coat, as the style she has in mind is what Poppy describes as 'completely naff'. Luckily for her, her skin will help her out. It will respond to the cold by shutting down all the blood vessels in its dermis to conserve heat internally. Consequently, her skin will turn white. She will also develop goosebumps and experience the evolutionarily useless reflex of her body hair standing up on end in a fruitless attempt to trap warm air. This is how, in the winter, our skin keeps us warm, and in the summer, because we can sweat, it keeps us cool.

Our skin responds to the ultraviolet light in the sun by providing us with a tan, which protects us to a very small extent from cancer and produces the so-called sunshine vitamin, vitamin D. This vitamin helps regulate the amount of calcium and phosphate in our bodies and is essential for maintaining strong teeth and bones. Lack of it can lead to deformities such as the bowed legs of rickets in children and the bone pain and tenderness seen in osteomalacia in adults. There has been a recent resurgence in the number of cases of rickets diagnosed in the UK as a result of vitamin D deficiency and the underlying cause is inadequate exposure to natural sunlight. It's true that too much sun can be harmful and that episodes of sunburn especially at a young age can increase the risk of developing skin cancer. But about 90 per cent of our vitamin D comes from the action of sunlight on our skin, even though the amount of exposure can be fairly minimal. Even with Poppy's dark skin, just 10–15 minutes of direct sunlight each day on exposed areas like the face, arms and legs is sufficient and is certainly not enough to cause burning or skin damage.

In quantifiable terms, how much does an average person need? Opinions vary, but vitamin D levels can be measured in the blood, and the general consensus is that when 25-hydroxy vitamin D falls below 30 nanograms per millilitre (nglml) in the blood, a person is deficient. But currently there is no standard definition of optimal levels of vitamin D, and it varies enormously between individuals. Eve, obsessed with healthy eating during her pregnancy, has been taking a 10mcg supplement ever since her Clearblue test was positive. She also insists her parents take one every day as she knows that the elderly are particularly at risk of deficiency.

Vitamin D may also be derived from our diet. Good sources include oily fish like sardines and salmon, eggs, fortified spreads and margarines, fortified breakfast cereals and powdered milk. Vitamin D isn't just vital for healthy bones and teeth, and Eve knows this. She was fascinated to learn that the highest prevalence of multiple sclerosis is found in countries furthest from the equator where the climate is cooler and sun exposure is minimal. In Scotland, for example, where there is a very low consumption of vitamin D-rich oily fish as well as a lack of sunlight, there is a significantly higher rate of multiple sclerosis than further south, and some authorities there are calling for the Scottish parliament to introduce mass supplementation.

Eve, partly because of the nature of her television work, has done her research carefully and doesn't get carried away by scary headlines. She knows the exact link between vitamin D and multiple sclerosis has not yet been unravelled, but it's an interesting theory. Perhaps a gene responsible for multiple sclerosis is affected by levels of vitamin D, rather than the vitamin D itself. At least, that was the suggestion made by one of the latest scientific studies. But she's not going to take any chances with her baby, and if there is even a remote possibility that deficiency of the sunshine vitamin in early childhood and in utero could increase the risk of that child going on to develop MS, she will be taking her supplements religiously. Especially since other studies have proposed that vitamin D triggers and arms the immune system, generally boosting levels of so-called 'killer' T-cells which react to and fight off pathogenic harmful bacteria and those more inconvenient viruses which cause the common cold.

As Eve rubs a thin layer of eczema cream over the flexures of Ruby's knees, elbows and wrists, they chat about the day ahead. What will Ruby have for lunch? Who will she sit with at school today? The skin-to-skin contact is nice, the bond between them reassuring and loving.

It's incredible how our bodies can change and adapt. Poppy will grow out of her acne. Ben's hair will become less greasy. Eve's stretch marks will gradually enlarge, while the vertical pigmented line between her navel and her pubic bone, and the areolae around her nipples, will darken once again during her pregnancy. Adam's skin will go enviably brown as he does the gardening in the summer, and will thicken and form calluses on the palms of his hands when he digs. Eve's parents' hair and skin will become even drier and thinner than it already is, so they'll increasingly feel the cold and develop more bruises when they knock themselves even gently, because they will gradually lose the collagen and other supporting structures in their dermis as a result of the ageing process.

The skin as an organ is nothing if not adaptogenic. We humans strive to adapt it even further. Our skin is perhaps the first thing other people notice about us after our eyes. Is it clear, healthy-looking and glowing? If not, why not? Cosmetically, in this day and age, good skin is everything. The interest in and take-up of botox injections and of collagen and dermal fillers is burgeoning. Chemical peels and dermabrasion and facelifts have never been more popular. People will spend fortunes in the pursuit of perfect skin.

And one day Eve hopes Ruby will have perfect skin too. It isn't so bad now, but when the eczema was at its peak, people in the supermarket or in the street would ignorantly stare. Obviously there is still a stigma attached to skin conditions. Bad enough to be bothered by red, scaly, itchy skin oneself, worse still to be sneered at by others. Eve has often had to physically restrain herself from reacting in public to these tactless, judgemental people.

How do people with more visible skin conditions cope when out and about and exposed to a critical public? People with alopecia areata who have lost whole clumps of hair as a result of an autoimmune disease? People with rosacea, with a butterfly distribution of a red rash over their cheeks and nose? People with vitiligo, where clearly demarcated areas of their skin have lost all pigmentation and turned white? People with keloid scars, where their scar tissue has become lumpy, firm and red as a result of an exaggerated healing process? People with sebaceous cysts, boils or impetigo, a common childhood infection caused by a simple bacterium? People with port-wine birthmarks or freckles? How much suspicion can still be aroused by the dermatological manifestations of ringworm, warts, scabies or cold sores? Folk would still recoil in horror, as they have done for centuries, to the Elephant Man, to lepers or to plague victims, thinking that they could easily catch the disease themselves. So how do people with third-degree burns on their faces or any exposed parts of their bodies learn to cope with the unwanted attention? It can be exceedingly difficult, and yet they do. They have to believe that deep inside they're just as beautiful and strong as they have always been. And that those who are taken aback, shocked and critical will never be as mentally tough and well adapted. They have to be, to coin a phrase, comfortable within their own skin.

Eve's thoughts return once again to Ruby and her improving eczema. There are certainly worse allergies to have. One of Michelle's children next door has a peanut allergy, and the other an allergy to sesame seeds. Both have been rushed off to hospital before now with anaphylactic shock. Michelle always carries an adrenaline-containing Epipen but even then it's a huge worry to her as she can never be quite sure when it might happen again and she cannot be with her children every minute of every day to protect them. What causes these potentially serious food allergies, and why do they seem to be becoming more common? How many people who claim to have food allergies genuinely *are* allergic, and what is the difference between real food allergy and food intolerance? Does food intolerance actually exist, or is it just a convenient and trendy complaint for hypochondriacs, people who aren't getting enough attention? It is easy for others to think so.

Food is something we all need and enjoy, and mostly it is one of our greatest pleasures in life. The variety of our menus and the creativity involved in the preparation and presentation seem endless, the multiplicity of textures and tastes an everlasting joy. So we rarely regard what we eat as a potential danger. Sure, we have all heard of the deadly deathcap mushroom Amanita phalloides and we know that accidental ingestion of poisonous deadly nightshade berries, Atropa belladonna, or of unsoaked, uncooked kidney beans or haricot beans can seriously harm us. But few people have stopped to consider that should they develop a severe allergic sensitivity to certain proteins in food, one bite of their innocuous-looking apple meringue pie or burger in a bun could prove fatal.

In nature, plants and animals have ways of protecting themselves. The Australian stonefish and stingrays of the type that killed the celebrated crocodile hunter Steve Irwin protect themselves with an incredibly powerful neurotoxin which can kill a man coming into contact with it in a matter of minutes. Venomous snakes and biting insects are capable of inflicting similar damage. Plants are no exception. Many varieties contain chemicals designed to ward off would-be diners as a form of self-preservation. These chemicals persist today, even in commonly grown crop plants. In some ways, in view of the enormous number of foreign proteins we consume in our diet, it is incredible that *more* allergic reactions don't occur. It is a tribute to our immune systems that this doesn't happen. One particularly clever type of chemical, however, is capable of turning the power of our immune system against itself. It fools usually protective immune cells in our body called mast cells into degranulating. This means the cell breaks up in the presence of the chemical and releases damaging inflammatory mediators such as histamine. It's this histamine that usually initiates an allergic reaction, from either an insect sting, pollen or foods. And this is why medications called antihistamines are the commonly chosen remedy.

True allergy to food is in fact still relatively rare. Up to 20 per cent of the population believe and claim they have a food allergy, but on scientific diagnostic testing fewer than one in ten of them genuinely do. That isn't to say everyone else is simply faking it. Their symptoms are real. It's just that the foods that upset them are not acting as a true allergen. They may be intolerant to these foods, and sensitive to them, but they're not by definition 'allergic'.

One group of researchers famously questioned 7,500 households, of whom 20 per cent reported at least one family member as having a food allergy. They tested them using a double-blind, placebo-controlled challenge test. This involved deliberately trying to provoke an allergic reaction in a cautious and controlled way by first getting them to eat a real food allergen and then a placebo or dummy substance, without either the person being tested or the tester knowing which was which. It turned out that only about 1.6 per cent of the adults actually had a genuine food allergy. The rest were obviously experiencing something else, such as a food intolerance, a food sensitivity or possibly even a food aversion.

A lot of people who say they are allergic to eggs or melted cheese simply don't like them. It's a ploy that many parents still use to permit their children to avoid having to eat certain foods they don't like at school. But if we use the excuse enough times, it could well be that eventually we begin to believe it. Some people, however, like Phoebe and Megan next door, *are* really allergic to certain foods. Highly allergic.

Many different foods can act as allergens, of which the most common for under-fives are cow's milk and eggs. Over the age of five peanuts, treenuts, fish and shellfish are more common. Phoebe's peanut allergy is potentially life-threatening. She doesn't just suffer from a blotchy red, itchy skin rash if she eats them: her immune system is so sensitized to even minute traces of peanut proteins that her lips and tongue swell, her respiratory airways shut down so that she cannot breathe, her heart races and she collapses and turns blue. Twice now, the adrenaline pen has saved her life.

So why does something as commonplace as a peanut provoke such a terrifying reaction? And why is this particular allergy becoming more common? Peanuts are an inexpensive source of protein and because of this they are used widely throughout the world, not just as a food but as an edible oil. This oil is used for cooking as well as in a wide range of cosmetic preparations such as shampoos and soaps. In the past it was also used in infant milk formulas and in medicinal creams and ointments. The allergens contained in peanuts are stable to heat and not inactivated by cooking, although good-quality peanut oil is free of allergenic proteins and therefore not a risk. Cold-pressed oils, however, may contain significant amounts of the protein. The ubiquitous nature of the peanut protein means that sensitization can often occur very early in life. As Eve is all too aware, it might even occur in utero.

Megan's sesame seed allergy is similar. Again, sesame is extensively used in the food, pharmaceutical and cosmetic industries. In the UK, about one in every 2,000 people are affected and, like Megan, they have to try to avoid any foods which contain sesame. Her mother has to be extremely careful to avoid vegetable oils which may contain sesame and all those lovely spreads that Megan would otherwise enjoy, such as humous and tahini. But just how potent can the peanut and sesame proteins be?

Peanuts contain very high concentrations of compounds called lectins, which are also present in other legumes such as beans and peas. You find them too in edible snails and wheat. The main feature of lectins is that they react with carbohydrate molecules on the surface of all human cells and change them. Red blood cells, for example, will clump together in the presence of lectins and this is how lectins are identified in the laboratory. They can also cause the breakdown of mast cells, triggering a catastrophic release of histamine and an anaphylactic reaction in people who are sensitized. For some people, lectins can be very bad news indeed. One type of lectin, ricin, is such a deadly poison it was used in the KGB's notorious murder of the Bulgarian dissident Georgi Markov. They had spiked the tip of an umbrella with a small quantity of ricin, and prodded him with it. It was sufficient to kill him. That was in 1978. More recently, in 2003 police raided a terrorist cell in a bedsit above a shop in London, where all the materials for manufacturing ricin on a large scale were discovered, including the beans from which the poison was being extracted.

Phoebe and Megan's parents are trying to protect them from a poison that is everywhere, a substance which so many of their little friends can enjoy with impunity, even in large daily quantities. It is a particularly selective poison. On top of that, because of the degree of their allergy and their age, they've been told to expect that it is not something they are likely to grow out of in the future. It is inconvenient, it is cruel and it is unfair. As far as Phoebe and Megan's parents are concerned, it is also personal.

It is vaguely irritating to them that other folk who clearly aren't affected in the same way as their daughters attribute much less severe symptoms to 'allergy'. Some have even assumed that Phoebe and Megan are just faddy eaters. There is so much confusion and misunderstanding about the subject. They have learned over the years that, amid all the charlatans making unsubstantiated claims about the value of unproven therapies such as applied kinesiology, cytotoxic tests, vega testing and hair analysis, real allergies are among the diseases least represented by specialists working within the National Health Service. Few people understand just how serious Phoebe and Megan's conditions are and how restrictive an effect they have on their everyday lives. Contrary to popular belief, they love going out to eat in restaurants and partying with their friends, it is just that they have to be extremely cautious.

As it happens, Eve herself is almost certainly intolerant to lactose, the sugar found in milk and other dairy products. Lactose is a natural sugar found in all milk from mammals and the enzyme lactase is made in our bodies in order to digest it. Insufficient production of lactase can result in people having lactose intolerance. This is least common among Caucasians and most common among populations in the Far East and Africa. Symptoms include stomach cramps, bloating and sometimes diarrhoea. Whenever Eve has been constipated as she is now in early pregnancy, she has often thought that any diarrhoea caused by lactose intolerance would be rather welcome. The more Eve talked about her irritable tummy and her food intolerances, the more she discovered that other people suffered too. One of her friends even has an intolerance to alcohol.

True allergy to alcohol is very rare indeed, although skin rash reactions have certainly been recorded. It can arise when the body is lacking an enzyme called alcohol dehydrogenase that's needed to properly break down the alcohol itself. More commonly, symptoms are caused by an intolerance to one or more of the ingredients on which the drink is based, for example, grapes for wine, grains for whisky or chemical additives such as sulphites. Alcohol also increases the permeability of the gut, which allows more undigested food molecules into the body. This might explain the reactions of mildly food-sensitive individuals who may not react to the food alone, but do when it is combined with alcohol. Eve is convinced that Adam doesn't have an intolerance to alcohol. On the contrary, as far as Eve is concerned, he seems able to tolerate it exceedingly well.

One type of food reaction many people *do* know about is coeliac disease. It's known that about 1 per cent of people in the UK have an intolerance to gliadin, part of the gluten protein that is mainly found in wheat, rye and barley. It tends to run in families and causes damage to the small intestine. Strict adherence to a gluten-free diet brings complete resolution of symptoms and this avoidance needs to be lifelong. Although coeliac disease involves the immune system, it is classed as an intolerance rather than an allergy as IgE antibodies, which are easily measurable in the blood and which react vigorously and specifically to true allergens, are not involved in the body's counter-attack. Interestingly, wheat or gluten intolerance are not the same thing as coeliac disease. You could have intolerance to wheat or gluten and not have coeliac disease. Studies have shown that the presence of IgG antibodies in blood can diagnose non-coeliac-disease gluten sensitivity.

Then there are chemical sensitivities. These are reactions

to chemicals in foods from colourings and flavourings such as monosodium glutamate and sunset yellow, caffeine, sulphites and tyramine. These reactions may occur quite quickly after any exposure, as small molecules can pass directly through the stomach wall into the bloodstream and then into the brain. This is thought to be why toddlers can behave in a hyperactive manner after being given brightly coloured sweets or fizzy drinks. But again, such reactions aren't strictly classed as 'allergies' as the immune system and IgE antibodies are not involved. Eve was amazed to read that people can also be allergic to histamine. The chemical that is released naturally by the body when you have an allergic reaction can also be found in foods such as fish, yeast extract, fermented foods and drinks, tomatoes, cheese and even some fruits and vegetables. All humans have an enzyme called diamine oxidase which breaks down any histamine from the histamine-containing food, but some people have a low level of this enzyme. When they eat too many histamine-rich foods they may suffer allergy-like symptoms such as headaches, rashes, itching, diarrhoea, vomiting or abdominal pain.

The diagnostic challenge with food intolerance is that the symptoms can be many and varied, and they can take several days to develop after the ingestion of the food. This makes identification of the culprit food and its elimination from somebody's diet very difficult. It partly explains why so many people guess at what might be causing their symptoms and avoid foods which aren't actually doing them any harm. Some people take this to extremes and end up with nutritional deficiencies as a result. It's one of the reasons why anybody suspicious of a food intolerance should always consult a qualified nutritional therapist.

But why has food intolerance become so much more common in recent years? Why are 45 per cent of the population reacting to foods their grandparents could seemingly eat with no problem? There are many different theories, but the main one is that we are eating less of the things we were designed to eat. Over the last 20,000 years, people have changed very little, especially genetically, but our environment and our food intake definitely have. It's likely that food intolerance has increased because of the poor quality of the soil, because of the use of chemical fertilizers and insecticides, because of more intensive rearing of animals, because we eat more processed foods and less fruit and vegetables, nuts and seeds. Also, there is environmental and chemical pollution and the increased use of medicines such as painkillers and antibiotics, all coupled with stressful lifestyles.

The key to understanding food intolerance probably lies within the gut. After all, the gut houses 70 per cent of our immune system and serves as a highly selective barrier against potentially allergenic food components and infections. At the same time, our gut lining selectively allows the passage of important vitamins, minerals, properly digested food proteins, carbohydrates and fats into the bloodstream. The trouble starts if the gut lining becomes stressed or damaged, when it can become more permeable, especially if alcohol is taken at the same time. Some scientists have called this 'leaky gut syndrome' or gut dysbiosis, where the cells which line the gut let substances through into the bloodstream when they shouldn't. It is in this situation that our bodies may start reacting to some of the foods we eat.

Eve has certainly done her research on food intolerance. Her own lactose intolerance had already got her attention, but a documentary she made on the subject drew her in even more. She was amazed at just how little knowledge about the new science of food allergy and intolerance existed within the orthodox medical profession and just how suspiciously it was regarded. It was clear that thousands of people were simply not getting the help and guidance they needed. Among other things which Eve's documentary had high-lighted, one particular strategy for fast-tracking the elimination diet process impressed her.

The method was a simple home-to-laboratory finger-prick blood test, called Yorktest, designed to detect IgG antibodies to a variety of foods. The test had been subjected to more published clinical studies than any other test for non-specific potentially reversible food intolerance. It was clear that these IgG antibodies were biological markers which could show that a reaction to a specific food type had occurred. Raised levels of these food-specific antibodies correlated with gut permeability and inflammation, which can then lead to symptoms anywhere in the body. It explained the wide range of food intolerance symptoms seen among the general population.

The best news, Eve's documentary had concluded, was that the majority of food intolerances were not for life. After a period of elimination, the culprit foods could often be reintroduced with no ill effect. If only it could be the same for poor Phoebe and Megan next door, with their peanut and sesame seed allergies. At least they didn't suffer from some of the wackier and more exotic allergies that Eve's documentary had highlighted, such as aquagenic urticaria – allergy to water. Affecting only one in 23 million people throughout the world, victims are confined to their own homes and react not just to foods and drinks containing water but to their own sweat and tears as well. For them, bathing and showering is almost impossible and they can only consume distilled water, free of the electrically charged particles called ions to which they are sensitive. A few hundred people in the world have been diagnosed as allergic to the sun. Solar urticaria is a mild form, but there's a much more serious condition called erythropoietic proto-porphyria where people's immune systems activate inflammatory cells within their skin when they are exposed to direct sunlight, resulting in diffuse and severe swellings.

Other people are genuinely allergic to exercise. Not in the way Poppy *says* she is, when she would just rather lie on her bed and listen to music, but in the sense that going to the gym could bring on a life-threatening anaphylactic reaction. Known as exercise-induced anaphylaxis, it only happens when some-one has eaten a certain kind of food such as peanuts first. The peanuts themselves don't cause a problem. But add on the exercise and the person will feel faint and nauseous and experience difficulty breathing. Then there's allergy to cold temperatures, caffeine, plastics and wood. A chemical called triphenyl phosphate used as a flame retardant in computers can promote electromagnetic hypersensitivity and the leather or dyes used in shoes can produce contact dermatitis. Allergies to the metal fasteners in underwear, to jewellery, to paper money or nickel coins are not uncommon either.

But the items in Eve's documentary that attracted the most attention and which were inevitably used for publicity purposes concerned allergies to kissing and to sex. Not excuses, the programme had pointed out, for avoiding contact with a partner you don't find attractive, but a real and genuine reaction to chemicals or compounds transferred in the process. In someone who has a severe food or medicinal allergy, the documentary explained that kissing another person who has been exposed to these things and sharing even minute traces of their saliva could trigger an overwhelming adverse reaction. In other words the kiss of death, definitely requiring treatment with the kiss of life. Even making love could have its hazards, the programme had explained. Latex allergy from condoms, for example, could cause very uncomfortable reactions for any amorous couple. And even when not using them, human seminal plasma hypersensitivity can render a woman allergic to her male partner's sperm. But they say every cloud has a silver lining and Eve had discovered that human seminal plasma sensitivity is no exception. Its treatment is gradual desensitization. This requires the couple affected to have frequent sex. Not surprisingly, Eve's documentary on allergies had been a runaway success.

Chapter 21

Roll on 16 weeks, Eve is thinking as she straps Ruby into her child seat in the back of her car and prepares to take her to nursery. Sixteen weeks is supposed to be that magical milestone in pregnancy when all those unpleasant initial symptoms like persistent headaches, fatigue, nasal congestion, nausea, urinary frequency and breast tenderness suddenly disappear, to be replaced by what Eve's doctor describes whimsically as the 'somnolescent tranquillity of mid-trimester pregnancy'. What would *he* know, she reflects, as she pulls out on to the busy road. There isn't much that is tranquil about her life in any stage of pregnancy. Or somnolescent, for that matter. A good night's sleep would be a fine thing. Below the horizontal strap of her lap belt she strokes the small bulge which is pressing forward on the baggy dress she's wearing and knows that in just eight more weeks it will reach her navel. Already her breasts are more tender and firm due to the increased number of glands that will produce milk when the time comes. Much to

Adam's delight, her bust has increased from a 34B to a 36B.

By the end of her pregnancy her circulation will increase by as much as 25 per cent in volume. She is already looking flushed, and her heart is working harder to pump more blood to her growing baby. Yet while the volume of her blood and the total amount of circulatory fluid is significantly increased, the numbers of oxygen-carrying red blood cells do not match this increase. Consequently, even though there are more of them, the blood tests Eve will have, just like the tests that every other pregnant woman has, will indicate a mild degree of anaemia. Obstetric science has never adequately explained this, but what we do know is that the increase in the numbers of infection-fighting white blood cells, initially slight, accelerate dramatically during labour and in the first few days after delivery. This is in order to mop up any microbiological threats to the mother and baby and to diminish the risk of puerperal (post-partum) infection.

The amount of blood pumped out by Eve's heart will increase by 30–50 per cent and at term her uterus will be receiving one-fifth of her entire blood supply. Her heart rate will speed up from its usual 70bpm to nearer 90 and when she is actually in labour her cardiac output will rise a further 10 per cent. No doubt at that time, she is thinking, the doctors will all become very excited again by the heart murmur she developed last time around. What is this, they had wanted to know, as a crowd of them took it in turns to clap their cold stethoscopes to her heaving bosom, only to mutter phrases such as 'haemodynamic flow murmur', 'systolic ejection phenomenon' and 'aortic stenosis'. What are they talking about? What do they all mean? Why couldn't they warm up their stethoscopes beforehand?

How could she remember so vividly the conversation they had had about the working parts of a stethoscope, when, as far as she knew, she could have been at death's door? How did their medical banter go again?

'What is the most important part of your stethoscope?' the consultant had enquired of the students. The first had suggested it was the bell-like apparatus that was applied to the chest. Another had insisted it was the calibre of the tubing which conveyed the sound to the earpiece. Another had opted for the two earpieces as these were the items that transmitted the sound to the ears.

'Wrong on all counts,' the consultant had said with an air of practised self-satisfaction. 'The most important part of the stethoscope is the bit between your ears. It's the inter-pretation of the sounds you hear that really matters. The real question is, what does it mean?'

And on and on he had gone while Eve could only wait patiently to find out what on earth they were talking about. What on earth *did* it all mean? They'd talked about systolic murmurs which occurred when the heart was contracting and blood was being expelled from the ventricles, either back upwards into the smaller atria as a result of incompetence in the mitral or tricuspid valves, or possibly because of narrow-ing of the main aortic valve at the outlet of the heart. Or was it a diastolic murmur, occurring in between heartbeats when the ventricles were actually filling? Was tricuspid or mitral stenosis restricting the blood flow from the atria into the more muscular chambers of the ventricles? Looking back, Eve remembered that within five minutes of the conversation she had convinced herself that her baby was soon to become an orphan. Whatever her heart murmur had actually meant, however, it couldn't have been much because all the echocardiograms and other investigations that they booked her in for once she and the baby had been discharged from the hospital were cancelled when the murmur promptly disappeared within a few days of her delivery. With no murmur to listen to, the doctors simply lost interest in her. Funny how just about everybody, including her friends and family, soon lost interest in her pregnancy once the baby was born, Eve thought. But wasn't that always the way?

A murmur, she had subsequently found out for herself, is simply the sound of turbulent blood flow. It's a rushing, swishing sound usually only heard through a doctor's stethoscope and is similar to the sound of water coursing through a kinked hosepipe at full pressure. In some people, it's a sign of a narrowed or stenotic heart valve or diseased artery. In children, it can be a sign of a congenital hole in the heart or just a vigorous circulation. In pregnancy, as in Eve's case, it is often just a manifestation of the increased blood flow and volume. It can also affect the elderly. Eve's mother, Grace, had developed a narrowing of the aortic valve at the outlet of her heart some years ago and had become tired, listless and breathless. Formerly active and energetic, she had become a shadow of her former self, had aged visibly and lost all her confidence. She hadn't been fit enough to withstand open heart surgery but had bravely accepted something called a TAVI procedure, which stands for Transcatheter Aortic Valve Implantation, whereby an artificial bioprosthetic valve is fed into the main artery from the heart through a catheter inserted at her groin. Crimped up tightly on the tip of the catheter, the artificial valve is attached to a metal coil, carefully positioned over the diseased valve and sprung open with sufficient force to compress the old valve against the artery wall and allow the new valve to take over its function. The result, for Grace, had been amazing. A new lease of life. More energy. Improved appetite. A zest for travel and gardening. And a photo and article in the local paper extolling the virtues of TAVI and the work of the local hospital. Joe, Eve's father, reckoned it had taken ten years off her apparent age. Eve's murmur by comparison was rather run-of-the-mill.

The circulation in her growing womb is changing too. Its increasing size is impeding the return of venous blood from her legs and pelvic area to her heart, and she realizes that, along with puffiness of her ankles and feet, her varicose veins will only get worse. Can she bring herself to wear those dreadful flesh-coloured surgical compression stockings from toe to groin to minimize the fluid retention? Not just yet anyway. Not a good look, as Adam was frequently reminding her.

At least, so far she hadn't noticed any problems with bladder infections. Increased oestrogen levels, which are changing so many parts of her body in so many different ways, are relaxing the tissues around her urethra, or water pipe, and encouraging germs to pass upwards from below into her bladder. Adam probably wouldn't understand; men, on the whole, don't experience this problem. A woman's urethra is relatively short, only about 4cm long, so micro-organisms can easily track upwards and infect the lining of the bladder, leading to cystitis. The copious amount of tea she is drinking is no doubt helping, however, as frequently emptying the bladder dilutes any germs lurking around and flushes them away at the same time. Besides, Eve's kidneys are working harder than normal now she is pregnant. Just like her heart and circulation, they are doing more and filtering an increased volume of blood. Between 16 and 24 weeks of pregnancy, her kidneys will be working to their maximum and only just before and during the birth will they slow down as pressure from the enlarging uterus decreases their blood supply. In a few months' time, Eve's kidney activity will temporarily increase if she sleeps lying on her left side rather than her right. This is because pressure from the enlarging womb is taken off the main vein carrying venous blood back to the heart from the legs, the inferior vena cava. So blood flow increases and the kidneys can work un-impeded. This substantial vein tends to lie more towards the right side than the left. At the same time, however, the enlarged uterus increasingly compresses the bladder, which because of its decreased capacity fills with urine much more quickly and will make Eve feel the need to empty it more frequently and urgently.

Inside the womb, although she can't feel it yet, Eve's baby is moving and turning, its tiny heart beating at least twice as fast as her own. In just a few more days, the midwife will be able to play out the sound of it using an ultrasound detector. The baby is well formed now with fingers and toes, the beginnings of nails and a large forehead and little button nose. The skeleton is complete yet still mainly in the form of cartilage. Some of it will have changed into bone by the time of the birth, but other parts will not fully ossify or become bone until puberty. The placenta is neither Eve's nor her baby's but shared. Once it completely surrounded the embryo and the amniotic sac but now it's fixed in position on one part of the uterine wall, while the chorionic membrane now lines the cavity of the uterus enclosing the amniotic fluid.

Neither Eve nor Adam has told many people about the pregnancy yet. The children know and are secretly thrilled, but otherwise it is still early days. While Eve's previous pregnancies were devoid of any serious problems, she knows she can take nothing for granted, particularly as miscarriage in the first 16 weeks is more likely in 'older mothers', as her doctor rather indelicately described her. An older multi-gravida sounded brutal, even if medically accurate. Her favourable previous obstetric history, however, is making her think hard about the prospect of a wonderfully orchestrated home birth courtesy of Mia, a long-time friend of hers who is a community midwife with a passion for home deliveries.

It had been tempting to tell the whole world as soon as the Clearblue test had been positive. Eve! Pregnant! Again! At her age? What was she thinking of? How marvellous! She could imagine the reactions. And the gossip. For that reason alone, she decided to keep the news private a little longer. She would wait for the first scan, wait for the first 16 weeks to pass, and wait to see how the land lay at work before telling her boss. Quite a few other women in the office had disappeared on maternity leave lately, and she wasn't sure what the reaction would be. It would be a shock for people to hear the news, but in many ways Eve was already desperate to tell them. She was surprised just how calmly she herself had taken it. It wasn't planned and she had always known the pill could fail, yet deep in her heart she is delighted to be expecting another baby. She loves her work and is good at it, but she has also loved being a mother, and now Ben and Poppy are almost adults it won't be long before they leave home and start to make their own lives. No. Eve is happy to be expecting another baby, despite the nausea and breast tenderness. Maybe that somnolescent tranquillity of pregnancy is just beginning to kick in after all.

Chapter 22

'DAD, CAN YOU just drive on a bit and drop us off around the corner?' asks Poppy. Like most of her friends, she doesn't like to be seen being dropped off by either of her parents. She knows her peer group, many of whom make their own way to school, will be watching. It makes her feel immature and over-dependent. A bit like Ruby being taken by the hand into the nursery. At 15, it just isn't cool.

Adam on the other hand is stressed enough as it is, as well as a little hungover, and this additional unnecessary request just adds to it. He obliges, but as he pulls out again into the traffic without so much as a backward glance, a pick-up truck coming up too fast behind him slams on its brakes and skids dramatically as it swerves around him. It must have been travelling well over the speed limit of 20 miles an hour, which has been set deliberately low outside the busy school, but as far as the driver is concerned, his inattention and reckless speed were irrelevant. Horn blaring and red-faced, he stops, gesticulates rudely and yells a string of expletives.

Totally oblivious to the teenagers and children around him, he is working himself up into a foul-mouthed F-word frenzy. Unfazed, Adam feigns ignorance, knowing for sure that his studied insouciance will make the driver's anger and need for retribution even worse. Ben is giving the pick-up driver as good as he is getting, if not actually outdoing him on the expletive and gesticulation front, and Poppy is even more mortified that now everyone within 50 yards will know that she gets dropped off in the mornings by Daddy. Who apparently drives like a cretin.

It's only when a community support officer with his self-important air and visible uniformed presence pitches up that the driver begins to pipe down and attempts to be polite. In the end, it is all a case of 'handbags at dawn', and Poppy and Ben amble off to join their respective pals loitering outside their schools, leaving Adam to continue his journey to work wondering what road rage is really all about.

Personally, he has never had a problem dealing with it. It has become more common, of that he is certain, but even though he admittedly failed to look in his rear-view mirror before he pulled away from the kerb, the pick-up driver's reaction was completely over the top. Had he given Adam the opportunity, Adam would have held up his hand in acknowledgement of his mistake and apologized. As it was, the man's verbal abuse, gestures and threats had only made Adam stand his ground. That, in a nutshell, is the basis of road rage.

Defined as 'aggressive or angry behaviour by a driver of a motor vehicle', it can lead to altercations, fights and collisions resulting in serious injuries and even death. The term was coined in California in the 1980s when a mounting spate of highway shootings prompted a response from the motor clubs, who instructed motorists in how to deal with drivers with road rage. Common manifestations were documented as sudden acceleration, braking and tailgating; chasing other motorists and cutting them up across lanes; and flashing their lights and sounding their horns. Obscene gestures became ever more explicit and creative at the same time. 'Vehicular homicide' became an inevitable legal entity. By 1997, psychologists and counsellors who had struggled to register road rage formally as a recognizable mental disorder nevertheless described it as a typical manifestation of 'intermittent explosive disorder'. And now that intermittent explosive disorder, as Adam knows only too well, is rife across the UK also. Commonly perpetrated by monsters in pick-up trucks, he thinks.

But the community support officer is on to it. Warming to his task and revelling in his power to quell the rage still boiling inside the driver of the pick-up, he calmly points out, in that rather clipped holier-than-thou delivery that all policemen favour, that road rage can result in criminal prosecution for assault under the Public Order Act of 1986. 'Under section 4a and 5 of the Act, you are causing harassment, alarm and distress,' he says. He seems to know it all by heart. 'And section 4 prohibits threatening, abusive or insulting words or behaviour with intent to cause that gentleman over there to believe that violence might be used against himself or one of his children.'

The pick-up driver is now clearly incandescent with a different kind of rage altogether. Authority rage and jobsworth rage. Two whole new categories to accompany air rage, road rage and computer rage. Not to mention 'roid rage, the unpredictably aggressive behaviour seen in bodybuilders taking excessive amounts of anabolic steroids. But the driver is not rising to it. He cannot afford to. The community support officer doesn't know it, but the nine penalty points currently recorded on his driver's licence are inhibiting him from reacting and risking further admonishment. So with enormous and commendable self-restraint, he holds it all in, his rage temporarily under control and a passive, repentant expression on his face as he endures this long, tedious disparagement of his driving etiquette. As he sucks it all up, Adam, with a huge and manifestly false grin on his face, steers neatly around the stationary pick-up truck, winks at the fuming driver and disappears off down the road.

Emotions are funny things though. Usually, they are short-lived responses to activities, thoughts or social situations that we experience every day. They are mainly unconscious reactions to opportunities or threats, but can also be consciously experienced as strong and overwhelming feelings that enrich our lives and give them value and meaning. They are generated in a part of the brain called the limbic system, which is not itself involved with consciousness. Strong emotions trigger associated activity in the cerebral cortex, particularly in the frontal lobes, which we experience as a conscious feeling or mood. Sometimes our emotions are obviously linked to something we are involved in, and sometimes they are not. But being aware of our emotions, being aware, say, that we feel sadness or anger or pity makes it easier for us to comprehend what is going on around us.

For example, when Adam drove past the scene of Toby's accident he felt a variety of emotions. A tinge of anxiety. A little sadness. A feeling of relief. And as the idiot in the pick-up yelled abuse at him in front of all these people, he felt embarrassment, anger and, momentarily at any rate, humiliation. All those signals coming in from the environment around him were fed to his brain and sorted for emotional content. One particular area was involved, the amygdala. This absorbs data from the sense organs and sensory cortices, and creates a neurological circuit through the cortex and hypothalamus. As his amygdala became operational, nerve impulses passed through his hypothalamus to

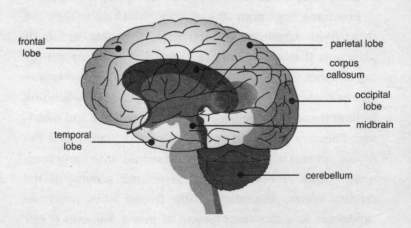

The Brain

bring about a variety of bodily changes and promote a conscious recognition of his emotions as they coursed through his frontal lobe. Negative emotions pass along one pathway and positive ones along another. On the right side of his cerebral hemisphere predominantly negative emotions are generated, followed by awareness of feelings like fear or sympathy, for example, which are perceived once signals have been passed across to and been processed by the left hemisphere. If these signals in the right side of Adam's brain hadn't actually reached the opposite hemisphere, his behaviour might still have been influenced by his emotions, although he might never have been conscious of them.

Moods are similar. Moods can be triggered very suddenly by things of which we may never be aware. And moods, unlike emotions, which are transient and fleeting, can last for hours, days or even months. It is the difference, in a way, between sadness and depression; between happiness and mania, and between a feeling of loss and of pathological grief. What emotions and moods have in common is the ability to bring about adaptive behaviour. Adam's sympathy for Toby made him subconsciously slow down as he drove past that fateful tree, and put him in an introspective and thoughtful frame of mind. His anger and irritation at being shouted at by the oik in a pick-up made him ready for a full-on confrontation or a fight. Rationally, Adam knew how to deal with the road rage situation very well. He'd practised the technique many times before. He realized that not rising physically to the challenge or threat would diminish further aggression in the perpetrator and tend to calm things down.

He also realized that provoking him verbally and making a joke of it all might do the opposite. Yet he felt powerless to resist the temptation. It was his way of not backing down and losing face. It was the human equivalent of locking horns, baring teeth and defending territory. It was the least aggression he could demonstrate to avoid becoming a victim.

In his mind, Adam had simply learned to regard the unnecessarily aggressive pick-up driver as a moron. Someone to feel pity for. A confused and agitated little man with a brain the size of a walnut who also probably suffered from erectile dysfunction and haemorrhoids. No wonder he was stressed out and abusive. Who wouldn't be? This was the technique Adam used to suppress his own anger. It was so successful he actually found himself feeling sorry for the guy. Not sorry enough to hope he wouldn't get three more points on his licence, but a little bit sorry. And as he drove on and neared his destination, he realized it would soon be his turn to confront the sources of stress in his own life, albeit of a totally different nature.

Chapter 23

BEN HAS NOW joined his regular group of friends while Poppy has wandered further up the road to congregate with her own at the school next door. Neither set of students is allowed into the school until 8.30, which means they are left hanging around in the playground chatting until the doors open. Ben has inevitably endured the ribbing he had expected after the unseemly fracas in the road.

'Has your dad actually *passed* his driving test?' asks one of his mates cynically.

'Should have gone to Specsavers if you ask me,' says another, to general merriment.

As it happens, Ben knows his dad has got an appointment at the eye specialist today but he isn't going to admit that right now.

'Maybe you could ask Toby to give him a few lessons,' at which even Johno, who is strangely quiet and introverted at the best of times, creases up with barely concealed laughter. The banter is cut-and-thrust, involving teasing, mocking and mimicry but all mixed up with elements of admiration and social solidarity too. There is a wariness, however, a consciousness of just how far any idea or conversation can be pushed, a probing of social standing and hierarchy within the group as each boy 'feels' how the others react and respond to him. A little good-humoured teasing of a general nature can bring the whole group together. Too much personal criticism, on the other hand, can lead to a full-on fight.

'Is it true your dad's name is Adam?' Joel asks Ben.

'And your mum's is Eve?' adds Paul.

'Adam and Eve!' they all chirp together with huge grins on their faces, relishing the delivery of a script they have obviously been rehearsing for some time.

'How come your parents didn't call you Cain then?' says Joel again. 'Or weren't they Abel?'

To which Ben, genuinely amused although he's heard it all several times before, smiles and nods repetitively as he stares at the floor.

'What have you got in your backpack today?' continues Tom.

'Is it forbidden fruit? From your mum's garden?'

'Very funny, guys,' says Ben eventually, mock-wrestling with the two boys nearest to him, but joining in with the fun at the same time. This physical reaction helps ease Ben's increasing discomfort and embarrassment, and allows him to deflect the mickey-taking from himself on to the more general topic of funny names.

'At least I'm not called Joe King,' he quips. 'Or Huw Jass.'

'Or Stan Still.'

'Or Gladys Friday.'

'Or Susan Orty-Boyden.' There is a few seconds' pause before they get that one and then they laugh out loud because of its deliciously school-matronly overtones. Johno takes a while longer and still isn't sure he really understands it. And finally, from Joel, the original perpetrator of the extended joke, 'Or Eve Ann Jellical.' Upon which Ben piles into him and knocks him to the ground.

Ben and his friends, like all other human beings, are extremely social creatures who need each other for mutual support, interaction and protection. The human brain has evolved and adapted in this way in order to make us more sensitive and responsive to one another's needs. To be able to live within an integrated group or community involves the ability to communicate with and understand other people and to be continuously aware of our social position in relation to them. To do this, we also have to possess a distinct consciousness of who we are in ourselves.

One of the human brain's most noticeable features is its very well developed outer layer of frontal cortex known as the neocortex. This is involved in the processing of conscious thought, emotion, reasoning, organization and planning. It is also vital to the requirements of learning to function within large close-knit groups. When Ben is messing around with his mates in the school playground, he has to adapt in just the same way as Adam and Eve have to do at work. They formulate appropriate responses based on other people's behaviour, they moderate their own, cooperate and respond to challenges. Innately we realize that other people hold beliefs and ideas which are different to our own and that they will see us in a certain light depending on the way we behave and act ourselves. The broad range and diversity of such social skills mean that many different parts of the brain are involved.

The amygdala, for example, unconsciously processes the significance of other people's facial expressions. When someone else smiles at us, we smile back. When someone else crosses their arms or even yawns, we tend to do the same. Whatever the expression on someone else's face, the amygdala analyses it for meaning. If it shows fear, we are likely to feel apprehensive ourselves. If it shows happiness and pleasure, we experience similar emotions and smile right back. Our natural tendency to mirror the movements of other people is inborn. This mirror neuron system, as it is called, is automatic. If the person talking to us scratches their nose, we often do the same thing. If they sit forward in their chair or put their hands in their pockets, we follow suit. And bizarrely, when the other person fails to mirror us, as we might expect them to do, it makes us feel uncomfortable, almost rejected. Brain scans carried out on people experiencing this can even highlight the distinct area of the brain responsible. It is the anterior cingulate cortex, an area which when activated can register physical pain from the body as well as the emotional pain of rejection. Then there is the insula. This is a brain region focused on the ability we have to separate ourselves from other people, to recognize the difference between 'us' and 'them', and the area which inextricably links consciousness to emotions.

Ben's discomfort at being mocked by a large group of boys is certainly not lost on most of them. This reaction is just part of the fun. If there was no detectable response, the game would not be worth playing. But each of the boys knows just when to stop. When enough is enough. When excessive teasing of any individual becomes bullying. Called the theory of mind, the ability to understand that other people think differently to ourselves and hold different beliefs and react differently is fundamental in all of them. All of them, that is, except Ben's friend Johno.

Johno finds it hard if not impossible to understand that other people feel differently about things than he does, and on top of that, he has problems with social skills and in communicating generally. Johno in fact has Asperger's syndrome, and for him, the theory of mind is absent. Johno sees mainly the facts of any given situation, not the reaction or interpretation of them in his friends. He copes with it, and is just as intelligent as the other boys in most other ways, but he finds it hard to maintain eye contact, can be extraordinarily blunt and rude without ever intending to be, and most of the other lads' jokes pass completely over his head. To those who meet him for the first time, he is hard going, humourless and expressionless. Yet to his friends, he is just Johno. He knows the school timetable by heart, not just for this term but for the next year, and the year after that. The only reason he doesn't know the timetable four years from now is because it hasn't been drawn up yet. Like many people with Asperger's syndrome, he has a penchant for facts and figures and a marvellous grasp of detail. Ben has already decided he will employ him as his accountant one day.

Ben's other friends are exhibiting a range of different facial expressions and body postures as he interacts with them right now. And his ability to read their various emotions enables him to predict to a certain extent what they will do next. It sounds complicated, a feat of inter-pretation that might take years to perfect if every individual exhibited different expressions, but analysis has shown that they don't. When human beings feel an emotion, a particular pattern of nerve signals is automatically triggered, which if we do not consciously subdue will bring about characteristic muscle movements in our face and body. Expressions and gestures in other words. Amazingly, how-ever, there is not an endless repertoire of these expressions in different people. There appear to be just a half-dozen or so main expressions which are more or less identical among different races and cultures and which are also the same in people who have been blind since birth. They are not copy-ing the facial expressions of other people, they are performing them innately.

Expressions are signals which we can read in others and which enable Ben to realize that his friends are teasing him with love, not malice. But these same expressions can be used to manipulate other people as well. Ben's current expression of helpless isolation and being an object of ridicule will soon be picked up by his mates (except Johno), who will quickly desist. And on her way to the nursery with Ruby, Eve's expression of curiosity and excitement is sending a strong message to her daughter that she is in for a lovely time today. So what are the half-dozen basic expressions which the entire population of the world seems to share? Happiness, sadness, fear, disgust, anger and surprise. All of which Ben has just clearly demonstrated in quick succession but not necessarily in that order. At this point in time, he's smiling. As are his friends. But what in essence does a smile consist of, and how do we know when one is genuine?

Smiling has probably evolved over tens of millions of years. Originally it was a sign of fear. But now even monkeys, apes and other primates use a facial expression of partially clenched teeth to demonstrate to would-be aggressors that they are in fact harmless and may even be enjoying themselves.

When it is born Eve's developing baby will almost certainly exhibit the trace of a smile within the first three days of its life, particularly when asleep. However, these earliest smiles are spontaneous reflex smiles and won't be a response to anything Eve or anyone else in the family is doing. The baby's first *social* smile won't happen for another two months or so, when it looks into Eve's eyes and hears the sound of her familiar voice. Another two months after that, and he or she will give their first genuine laugh.

A smile, we all know, is a facial expression formed when the muscles near the corners of the mouth contract. But when someone smiles, are their eyes also involved in the smile or is the smile purely confined to the lower part of the face? The distinction is important, as one conveys much more meaning than the other. Anyone can easily produce a social smile if they think about it consciously. We use such smiles to increase our likeability, attractiveness and friendliness towards others. We do it socially at dinner parties and other gatherings and we do it at work too. But an unforced smile anatomically uses an additional set of muscles situated around the eyes.

First described by the French neurologist Guillaume Duchenne, this smile signals the genuine emotions behind it, and really does come from the heart as well as the head. Hard to produce on demand because it needs emotion to fuel it, it is a sincere spontaneous manifestation of a joyous mood and a warm welcoming reaction to an event or another person. When it comes, it is radiant. The lower lids of the eyes puff up and the skin at the outer corners wrinkles up into crow's feet. The eyes themselves twinkle and sparkle. There is an almost imperceptible dilation of the pupils too. Here in the Duchenne smile, the signals coming from the brain emanate unconsciously from the amygdala, and pass through the motor cortex, from where messages are sent to the orbicularis oculi muscles of the eyes, and the zygomaticus minor and major muscles at the side of the mouth. In the social smile, learned as part of behavioural etiquette, signals emanate from conscious areas of the brain to premotor and motor cortises from where messages bypass the eyes altogether and only influence the muscles of the mouth. Just like facial expressions, a smile is almost universally recognized by people all over the world as something which can communicate affection, warmth and even sex appeal. Regarded positively almost everywhere, it is often regarded as a prelude to laughter or a response to somebody else's laughter. Research has shown that smiling engenders a greater sense of trust and an improved feeling of inter-personal cooperation. Even the smile of a stranger can promote 'good Samaritan' effects in the recipient.

But a smile can also be overdone or open to interpretation, and small but subtle cultural differences do exist. Many women might view a man they do not know who smiles persistently at them as threatening. Excessive smiling can also be perceived as a sign of shallowness or dishonesty. A fake smile can signify contempt or mockery, a genuine one pride or love. Some races smile when they are angry and confused, others use a smile to camouflage emotional pain or embarrassment.

As Shakespeare said, 'There's no art to tell the mind's con-struction in a face,' and that's certainly true when it comes to the social smile. You really can't tell a book by its cover, or a sausage by its skin. The genuine Duchenne article, however, tells a different story. That one is for real and comes straight from the heart. And that's the smile that Ben is sporting now, as he picks Joel up off the floor and dusts himself down. Johno is smiling too. But he doesn't really know why he's doing it. And his eyes haven't changed a bit, because Johno finds it difficult to understand people's emotions at all.

Chapter 24

NO MORE THAN 300 metres further down the road from Ben, Poppy is also surrounded by her own group of friends. Since they all stop talking suddenly when she joins them, she guesses that they'd probably been talking about her and wonders whether she should regard them as temporary acquaintances rather than true friends. Real friends don't talk about you behind your back. Girls interact differently to boys though. Boys can josh around, take something too far, get upset, have a fight, beat the hell out of each other and then forget all about it and make up a few minutes later. Girls are generally more canny and manipulative. They use guile and subtlety rather than muscle and brawn. Whoever coined the phrase 'sticks and stones may break my bones but words will never hurt me' was so utterly wrong. Persistently cruel verbal abuse from a quorum of girls can be torture. Especially to sensitive self-conscious souls like Poppy. Their barbs, thinly disguised put-downs and snide remarks can be as cutting and hurtful as the guillotine itself.

Whether directly to the face, through Chinese whispers or indirectly on Facebook or Twitter, girls are capable of bullying other girls in all sorts of ruthless ways and many seem to derive intense pleasure from it.

How many cases of self-harm, eating disorders or even suicide can realistically be attributed to constant bullying from a girl's own peer group? Poppy can only guess. At least she can always rely on Anna and Kirsten, her best and oldest friends. They had been at her previous school too and they had come through all kinds of adversity together. They also had to integrate with the current bunch of girls, the majority of whom had come from the college's feeder school and still regarded Poppy, Anna and Kirsten as outsiders. This sense of exclusion just made the trio socially tighter. They were discreet, supportive, loyal and genuinely very fond of one another. Nothing could come between them, and it was a bond that the other girls were clearly envious of.

They were a strange bunch. There was Carly, with her increasing collection of tacky-looking tattoos, Peta, who was boasting about yet another body piercing, and Sinead, who was so fat she looked like she'd been poured into her school uniform and had forgotten to say 'when'. There was Davina, with her hirsute upper lip and sideburns, Tess, with her unfortunate gluteal hypertrophy (the others just said she had a large bottom), and Maggie, with her particularly prominent chin. And then there were all the other ugly miscreants who had looked studiously down at the floor and kicked their heels in feigned innocence as soon as they'd seen Poppy approaching.

'My, my,' said Poppy aloud to herself suddenly, 'I seem to have picked up the bitching bug too.' Funny, she thought, how nice individuals can get into groups and pick up a hunting pack mentality. If she didn't watch out, she'd be joining a gang. As long as it's Kirsten's and Anna's and *her* gang though, she wouldn't really worry. As far as she was concerned, it was all about survival.

What makes people behave in such different ways? Poppy wonders. Why are some people sweet and nice and stay like that all their lives, whereas others are rude and aggressive? Why are some people totally extroverted and larger than life, while others are shrinking violets and all but invisible? Why do some people take risks or indulge in addictive behaviour? Why are others cooperative and helpful, altruistic and kind? What makes one person always inclined to look on the bright side of life when others are eternal pessimists?

Take Peta. She's the same age as Poppy, just 15. Yet she's been smoking cigarettes since she was 11, is regularly intoxicated with drink, has experimented with a variety of recreational drugs and just wants to drop out of school as soon as possible. Lesley sleeps around with boys because she has no sense of self-worth, Billie is obviously bulimic, Sasha hardly utters a word but will get straight As in her GCSEs and Fran is just Fran. She is the irresponsible, do anything, any time, any place, life and soul of the party. It's difficult for Poppy to decide whether any of her group are actually normal. But what is normal when it comes to personality? And how are our personalities developed and formed?

Poppy's personality is sociable but self-conscious.

Imaginative but slightly insecure. Bright and inquisitive yet down-to-earth and fond of routine. Calm but somewhat disorganized. Fun-loving but not excessively extroverted. Like all of us, she does not have just one personality trait, one behavioural characteristic that dominates all the rest. Her character is multi-faceted and capable of exhibiting different traits at different times and in different situations. There is the sympathetic, concerned Poppy when she is caring for one of her upset friends. There is the sad, self-loathing Poppy when her spots are bad. There is the laughing, impish Poppy when she is at parties receiving attention from the boys. And there is the thoughtful, anxious Poppy when her exams are approaching and the revision has to be done. None of us is just one thing. And our personalities can alter according to changing circumstances. Some people's personalities change markedly on a regular basis. Others change very little.

To an extent, all of us are genetically pre-programmed as to who we are, and who we will be. Organically, individual parts of our brains are responsible for how we behave. One day, it might be possible to map every cell in our brain and predict exactly how we will behave in any given situation. Neurologists know a great deal about this already. Optimists, for example, demonstrate much greater excitation in the anterior cingulate cortex and amygdala parts of the brain when contemplating positive future events compared to other people. Sociable people experience enhanced responses in the striatum region of the brain which is associated with reward and pleasure. Aggressive people have a smaller and less active cingulate cortex, another area of the brain usually charged with the monitoring and moderation of impulsive behaviour.

This might explain why people who exhibit extraordinary behaviour or who act in ways far beyond what is considered normal do what they do. Einstein's brain, for example, which was studied after his death, was found to be wider than average, and the groove which usually divides a section of the parietal lobe was not present. Since the region affected is associated with calculation and spatial reasoning, it could well be that nerve cells there were able to combine more effectively with one another, endowing him with his mathematical genius. Conversely, certain murderers and serial killers who were never demonstrably emotionally or socially damaged by their upbringing and environment may well have perpetrated their crimes as a result of some slight anatomical discrepancy in their brain.

Frontal lobe tumours can definitely alter people's personality and behaviour too and it is undoubtedly the case that more subtle neurological idiosyncrasies can work in the same way. Very recent research has examined the brain scans of antisocial people in great detail. Compared to a control group, they revealed an 18 per cent decrease in the volume of their brain's middle frontal gyrus and a 9 per cent reduction in volume of the orbital frontal gyrus, two sections of the brain's frontal lobe. Other brain studies carried out on psychopaths who use manipulation, violence and intimidation to control others and satisfy their own self-centred needs have shown an 18 per cent decrease in volume of the amygdala, neatly accounting for their total inability to feel sympathy, remorse and guilt. Is it possible that the mass murderer Fred West's actions could be explained by the metal plate he'd had inserted into his head following a motorcycle accident at the age of 17? Is it possible that the terrorist kidnapper and bomber Ulrike Meinhof behaved in the way she did as a result of the operation she had for a swollen blood vessel in her brain several years previously? All we can say is that small aberrations from normal can sometimes result in profound changes in behaviour and personality. But this is only half the story. Our genes give us the raw material from which our central nervous system develops, but it is our upbringing and environment which teach us the rest.

Personality can also be regarded as a series of familiar responses. These reactions can be learned through the mimicking of our parents' behaviour or that of our peer group or siblings. They can also be altered and manipulated and, depending on how impressionable we are, significantly influenced through powerful media such as TV and films. Why else would watching a violent film or playing electronic war games make some people more likely to commit violent offences themselves? The TV and film industry do not like to admit it, but we know that this happens.

The personalities of the girl group around Poppy are very varied. That's what makes people fascinating. The different family settings, cultures and environments will have made them who they are, and to a large extent they are still developing and changing. One day in the future many of them may well be subjected to psychometric testing prior to job interviews or possible promotions. These personality tests attempt to categorize people into groups with a predominance of certain behavioural traits or to profile them on where they stand along a spectrum of different characteristics. Poppy likes to do this job herself, however. She knows very well who among this group of girls she likes the best and why, and no tick-box piece of paper is ever going to convince her otherwise. Thankfully, her own personality makes her very tolerant of others. What business is it of hers that Peta now has 16 body piercings, and that Carly's tattoos are making her look increasingly like a lace curtain?

Piercings and tattoos are body decorations that go back centuries. They are designed to make a statement, attract interest and attention, separate people from the crowd and get a reaction. They work, but at what cost? Poppy can understand the beauty of a subtle, understated symbol placed somewhere discreet for private viewing only. But she feels her friends have become carried away with the rebellion of it all. It is frowned upon by their teachers and their parents alike, it breaks the college's rules and it gives a certain signal to members of the opposite sex about attitudes. For Peta and Carly it has become something of an obsession, a behaviour which consumes and defines them, an interest which Poppy considers has got out of hand and has made Poppy herself even more determined never to go down that route. And that's not to mention the fear of complications.

Peta has a number of red angry-looking mounds of flesh growing around her body jewellery. There is a glistening

excrescence of raw skin resembling tiny grapes around the
ring in her eyebrow, and the decoration in her belly button.
This is granuloma formation, which occurs when the body
attempts to heal itself over-vigorously in the presence of a
foreign body. Peta's had several treatments using a silver
nitrate cautery stick to burn it away, but because the
jewellery is still there and she steadfastly refuses to take it
out, it will just keep forming. She also has keloid scars –
raised and lumpy livid reactions to where an infection
occurred after she'd had 'dermals' placed in her buttocks and
at the top of her cleavage. These mock diamonds are pinned
through the skin into poppers like press-studs underneath,
but they aren't real diamonds, just a much less expensive
alternative. Unfortunately, the cheaper the materials, the
greater potential for allergy and rejection.

Poppy thinks the bother of having to put up with these
reactions to body piercing would be bad enough, but for her,
the main objection is the desecration of the natural beauty
of the human body. Her own adolescent acne apart, why
would anyone want to permanently scar, discolour or stain
their skin? Or all three? Why would anyone want to run the
risk of contracting hepatitis or AIDS at the hands of un-
qualified tattooists using unsterile needles in dirty premises?
Or risk permanent sensitization to chemicals in the dyes
used in tattoos performed abroad, so that any application of
commonly used hair colourants might result in horrendous
allergic reactions with blistering and oozing dermatitis of
the scalp and face? Why would anyone want to disfigure
themselves using cheap-looking mass-produced designs that

you could quickly tire of but never be able to remove? How will they look at the age of 40 when their skin begins to sag, when fashion has moved on, and anyone who is anyone will see them as ugly and distasteful?

Peta and Carly have all the answers. The more they are challenged on the subject, the more they rebel. Poppy reckons if they knew half as much about their GCSE subjects as they do about tattooing and body piercing, they'd both get straight A stars and qualify with distinctions. What was it Carly had told her just yesterday? Between 73 and 82 per cent of women in the US have had their ears pierced. The number of women with tattoos quadrupled between 1960 and 1980. And yes, it is imperative to only go to a recognized practitioner on a voluntary register employing proper infection control guidelines. And so on and so on. Poppy admits they're knowledgeable but, in her opinion at least, they are still misguided.

Peta especially is beginning to look like someone far too wacky and weird to even appear on *The Jeremy Kyle Show*. Poppy can't help feeling the way she does about the choices Carly and Peta make about their appearance, yet she knows only too well how irritating it can be when her own mother comments disparagingly about her own. She also recognizes that Peta, despite her image, can be one of the kindest and most generous people she knows, rarely given to bitching or talking behind someone's back like some of the other girls. The different characteristics and personalities in the group of girls surrounding Poppy are partly determined and hard-wired by genetics and heredity, but also partly random and influenced by environment, fashion and popular culture. Different as they are as individuals, as the bell sounds for the start of their school day, they turn uniformly and walk towards the school as one.

Chapter 25

AT EXACTLY THE same time as Eve is squeezing her car into the last remaining parking spot in the car park outside Ruby's nursery school, Adam is approaching the office where he works as the chief executive of a relationship counselling charity. He is uncharacteristically nervous and abnormally conscious of feeling rather stressed. The hangover doesn't help, and he's bitterly regretting the decision he made several months ago to agree to speak to a group of selected sixth-formers at a local school. He's due there later in the day to talk about sex education and relationship issues. An experimental initiative, it was brought about by a request from the pupils themselves as part of their pastoral and moral development and had been endorsed by the formidable principal, the Parents' and Teachers' Association and the school's board of governors. Adam had been flattered to receive the invitation. Who wouldn't be? It was a recognition of the significance of the charity's work, a tribute to his own communication skills and an interesting challenge.

His mandate was to talk about the importance of relationships, romance and sex, but the format would be a pioneering no-holds-barred any-question-tolerated free-for-all in the school assembly hall with 200 pupils and only two other adults present, neither of whom would be regular teachers. One was a school chaplain from a different establishment and the other was a Family Planning Association-trained nurse currently seconded to a government-backed pilot scheme providing contraceptive implants and intra-uterine contraceptive devices (coils) to schoolgirls as young as 13.

What on earth had he been thinking of when he agreed to do this? Adam wonders to himself. This had to be one of the trickiest and most sensitive subjects he could ever have imagined talking about to anyone, let alone a cynical bunch of immature spotty adolescents with raging hormones and rampant sex drives who would probably just want to get laid for the first time and have no regard for anyone else's feelings. He thinks he may have put his head in a very tight noose. Suddenly he feels quite sick and stops the car, opens the door, steps out and takes deep breaths of cool fresh air. His pulse is racing, his heart thumping and he feels a rush of cold sweat suddenly spring from all over his body. Never again, he is telling himself, will he agree to take on such challenges. Never again will he put himself in the firing line just because no other sucker was foolish enough to take it on and he was flattered to be asked.

By now Eve, not a million miles away as the crow flies from her ashen-faced, queasy, perspiring husband, is leading

Ruby into school. She'd got her there just in time before her recent car sickness manifested itself, and the practised distraction techniques she'd been using en route had clearly paid off. Coupled with her own nausea, Ruby *not* throwing up her recently consumed porridge was a blessing. What is it about travel that brings on such queasiness, she wonders, and why do children seem to suffer from it more than adults? She'd vaguely heard people talking about motion sickness, something called the labyrinth and the vomiting centre, but she'd always thought the labyrinth was a patented wooden game of coordination for children, and the vomiting centre sounded like a good place to visit after eating too much chocolate or seeing someone pick their nose.

Most people suffer from travel sickness at some stage of their lives, but unless you are a regular sailor on particularly challenging voyages, you will probably have grown out of it by your teens. It happens when the brain receives messages from the organs of balance in the inner ear, the labyrinth, and the eyes at the same time. If the messages are in conflict with one another, say when you look at the interior of a car which isn't moving relative to the rest of you while your other senses tell you you *are* moving, a feeling of nausea results. Children are more naturally susceptible to it anyway, but their short stature often means that they cannot see the horizon outside the car's windows and cannot therefore compensate for the conflicting messages.

Eve had tried sitting Ruby higher up on booster seats and preventing her from looking at pictures in books while she travelled, but neither had really helped. Nor had opening the windows, getting her to wear acupressure bands on her wrists or feeding her ginger biscuits, a much-vaunted natural remedy for motion sickness. Her neighbour's partner, Colin, a yachtsman, had rather unhelpfully suggested that Eve should give Ruby pineapple for breakfast. He had recently read an article about it in *Yachting Weekly*, where research trials had compared various anti-sickness medications with a number of natural alternatives. Of all things, pineapple had come out on top. It didn't help prevent sickness, he'd said, but it was the only thing you could eat that tasted the same coming up as it did when you swallowed it. It was a fact that Eve, hard though she tried, was finding very difficult to forget.

Once in her classroom and absorbed by the paper and crayons laid out on her desk, Ruby never once looked back. Strangely, it was Eve who felt a pang of separation anxiety today, thinking just how quickly the five years of Ruby's life have whizzed by. Where have they gone? she wonders. It seems like only yesterday that Ruby began walking, yet that was four years ago when she was a one-year-old. At two, she was pointing to different parts of her body, eating tidily with a spoon and fork and had become dry during the day. Another year later, she was talking in full sentences, naming colours and dressing without help. Her growth had been remarkable. Between the ages of one and three, Ruby had gained an amazing 40 per cent in height and weight, growing faster than she would at any other stage of her life. It was the equivalent, Eve knew, of a 12 stone, 6-foot adult growing to 16 stone 11lb and 7 foot 8 inches tall in just two short years. Incredible. During that time, there was significant development of her brain, liver, heart and other organs. Eve had been aware of all this and, like any good mother, had done her utmost to ensure the best possible diet for her daughter.

The nutritional requirements of a toddler of this age are very different from those of older children and adults. A toddler needs 4.7 times the amount of iron per kilo body weight compared to an adult, 2.8 times the amount of energy-giving foods, 2.9 times the amount of calcium and 4.4 times the amount of vitamin C. Eve had known not to give Ruby skimmed or semi-skimmed milk at that time, or any other low-fat dairy product, which would be more suitable for an adult. She had made sure that she included all of the five food groups in the right amounts. She had seen to it that Ruby ate them as three meals supplemented by two satisfying and refreshing snacks every day. She had given her bread, rice, pasta, potatoes and other starchy foods with each meal, together with a variety of fruit and vegetables. Ruby had had three servings of milk a day, including cheese or yoghurt, and two or three servings of fish, eggs, nuts or pulses, with only small amounts of the foods and drinks that so many other toddlers in her class seemed to be given, which contain such high levels of fat and sugar. Eve had realized how difficult it was to ensure that toddlers got all the nutrition they needed. She had known how many things could get in the way. Food refusal, faddy eating, teething, illness, tiredness, the beginning of nursery and children's parties. Eve is aware that many toddlers today have anaemia or inadequate vitamin D and omega 3 and 6 essential fatty acids in their diets. She also knows that this is part of the cause of so many behavioural and learning problems such as dyslexia, dyspraxia, attention deficit hyperactivity disorder and autism.

That's why she currently insists her family all spend much more time sitting down to eat together. She cooks as much food from scratch as she can and tries not to rely too much on pre-prepared dishes designed for adults, which are often very rich in salt and sugar. She keeps chocolates and sweets for occasional treats rather than routine events and she does the same with crisps and fizzy drinks. Eve is always encouraging her children to eat lots of leafy green vegetables and fruit. She had started them at a young age and had patiently reintroduced the foods that at first they didn't seem to like. She is determined to do the same with the next baby. Eve has done her best, yet even she realizes that with the best will in the world and probably a first-class degree in infant nutrition, it is impossible to ensure a perfect diet 100 per cent of the time.

At the age of four, Ruby was bouncing and catching balls from the floor. She could hop on one leg and draw an impressively lifelike picture of her older brother and sister. Now aged five, she is becoming increasingly grown up and independent.

In the car on the way to school, Eve had got Ruby to play Guess What? What's got black and white stripes and four legs? she'd asked. Ruby shot right back with 'zebra'. She'd given her a set of commands to follow. Touch your cheek.

Now your nose. Now your chin. And Ruby had obeyed every one. Eve had got Ruby to mentally put things into categories and Ruby had known that teddy bear, doll and dominoes were all different sorts of toys. She could name ten objects beginning with R, starting with Rufus of course, and ten more R-things she could do with Rufus like run with Rufus, race with Rufus and read with Rufus. Eve had encouraged Ruby to ask more of those unanswerable questions which only children with a highly developed sense of curiosity seem to ask. And Ruby had responded with interest.

Why does Daddy leave his expensive car on the drive outside the house and lock up all his junk in the garage? Why don't fortune-tellers ever win the National Lottery? If mosquitoes are so horrible and spread such horrid diseases, why didn't Noah swat them before they got on the Ark? And lastly and rather politically incorrectly, thought Eve, why do very fat people eat double cheeseburgers with loads of chips and fried Mars bars, and then order a Diet Coke? Ruby's questions were thoughtful and insightful, which amazed Eve and proved her daughter is bright and imaginative. Her cognitive development is coming on in leaps and bounds and her behaviour and daily routines are changing rapidly. She is forever expanding her horizons and inevitably spending more time away from the familiarity and comfort of home.

Yes, she still needs cuddles from Eve and Adam and she still occasionally displays emotional extremes with tantrums, sulks and moodiness. Increasingly, though, she is forming relationships with her own little clique of friends and gravitating towards other children of a similar age. She is also developing a very good idea of the moral difference between right and wrong. She now understands the concept of rules and regulations, despite wanting to challenge them and push the boundaries whenever she feels like it. She has become curious about relationships between any adults she comes across, and she enjoys giving, sharing and receiving in equal measure. More and more, Eve wistfully acknowledges, she is developing a tendency to want to spend time on her own. In her head, Eve knows it is normal and even healthy, yet she can't help feeling that pang of maternal redundancy.

Chapter 26

ADAM MEANWHILE, STILL leaning against the side of his car
and breathing deeply, is all too familiar with the signs
and symptoms of a hangover. His head is pounding, he feels
weak and queasy, but most of all, he simply feels that he
hasn't slept for a week. Any bright light hurts his eyes, everything seems louder and more intrusive to his ears, his
muscles ache, his eyelids are red, and he cannot shake off this
unquenchable thirst. On top of that, his heart keeps racing,
he feels dizzy as if the world is spinning around him, and his
hands are clammy and slightly shaky.

How on earth is he going to get through the day when all
he really wants to do is put his head down and sleep it off?
Today of all days, he is thinking. Judgement day, when 200
teenage pupils will be hanging on his every word in an
experimental and inaugural masterclass on anything
remotely to do with sex. Maybe it won't turn out to be
inaugural. Maybe it will be a one-off. Maybe he'll make such
a pig's ear of it neither he nor anyone else will be invited to

speak ever again. Maybe the repercussions of holding such a transparent and honest general forum on the topic of sex education will tarnish his own and the school's reputation for years and unwittingly even set back ambitions to stem the rise in unwanted teenage pregnancies, sexually transmitted infections and under-age sex?

Just thinking about it makes his head throb even harder. Hopefully, the hangover cure he has perfected over the years will kick in soon and revive him. He's already swallowed his two Nuromol tablets with a hastily concocted banana milkshake made with full-fat milk, cream and honey at breakfast time. The caffeine in the can of Red Bull he is drinking now and in the large heavily sugared latte he is about to purchase should do the rest. Over the years, Adam has read many a scholarly article on the subject of hangover cures and only two other solutions have proved even remotely effective: a hair of the dog, and lying face-down in a snowdrift. Neither of these was going to be practical today, however.

The physiological effects of an alcohol binge are well documented. The direct effects include dehydration, electrolyte imbalance, gastrointestinal disturbance, low blood sugar and biological rhythm alterations. Then there's the effects of alcohol withdrawal, of alcohol metabolism, and the non-alcohol effects of compounds other than the alcohol in beverages such as methanol and all the unhealthy food often consumed at the same time.

Alcohol also makes the kidneys produce more urine. It is a diuretic. Adam would have noticed this the previous night when he was consuming several pints of beer and rushing off to the gents' every 15 minutes to empty his bladder. This activity, which many people attribute to being over-hydrated, is paradoxically a function that ultimately desiccates you like a dried prune. Diuresis means you pee more at the time that you're drinking, allowing more water than usual to be lost, resulting in dehydration unless water is drunk to compensate. The effect is measurable. Fifty grams of alcohol, the same amount that might be found in say four drinks, causes the elimination of 600ml to 1,000ml of water over several hours. Alcohol does this by inhibiting the release of the anti-diuretic hormone (ADH) otherwise known as vasopressin from the pituitary gland at the base of the brain. Reduced levels of ADH prevent the kidneys from conserving water and therefore increase urine production. Further dehydration can occur through sweating, vomiting or diarrhoea, but Adam, thankfully, has at least been spared that. What he hasn't escaped, however, are the irritational effects of alcohol on the lining of his stomach. Gastritis, as it is known, is particularly common when stronger drinks with a higher alcohol concentration are consumed because these delay stomach emptying. So the single malt whiskies he'd downed towards the end of the evening were a mistake. In addition, increased secretion of intestinal and pancreatic juices can contribute to the twinges of upper abdominal pain often experienced by hangover sufferers the morning after the night before.

Adam's blood sugar is still on the low side, as he queues for what seems an inordinate length of time for his coffee. Why is there only one guy making each individual coffee from scratch at 9am when the world and his wife want to start the day with a caffeine-charged eye-opener? he is thinking. Maybe he should write to the café's management team with suggestions. Maybe his current feeling of exhaustion means he won't bother. His low blood sugar levels are making him feel both hungry and drained at the same time. Several alterations in the metabolism of his liver have brought this about. Fatty changes in the cells of his liver and an accumulation of lactic acid have combined to inhibit the production of glucose in his body. Glucose, however, is the primary energy source for Adam's brain. He's going to need that glucose today, yet the lack of it at this moment is making him feel weak, moody and tired. In fact, cripplingly and overwhelmingly tired. How can it be that having a few drinks at a party makes you feel so sleepy when you go to bed and puts you in such a deep coma for the rest of the night that you wake up feeling as if you haven't slept at all?

Despite the short-term sedative effect of alcohol, it is in fact very disruptive to normal sleeping patterns. The initial sedating effect of the small amount of alcohol in a nightcap soon wears off, leading to a rebound alertness of the brain soon after falling asleep, causing insomnia. Larger amounts of alcohol can render someone comatose for hours, yet still leave them sleepy and tired all the following day. This is because the normal sleep pattern is turned on its head, so that less time is spent in rapid eye movement sleep and much more time in deep slow-wave sleep. Because alcohol is a powerful muscle relaxant, the muscles at the back of the throat relax a lot more, so the airway can become more easily obstructed leading to exaggerated snoring and sleep apnoea. The latter, which involves the occasional cessation of breathing altogether when the airway becomes completely blocked, will wake the person who experiences it regularly throughout the night, and seriously interfere with both the quality and the quantity of their rest. Alcohol also interferes with normal 24-hour biological or circadian rhythms. Body temperature, for example, becomes lower during intoxication and abnormally high during a hangover. Night-time secretion of growth hormone is disrupted, affecting protein synthesis and bone growth. Normal fluctuations in levels of the stress hormone cortisol are also seen.

If all these changes were not enough of a 'headache', for Adam his physical headache caps it off. The alcohol in his bloodstream has dilated blood vessels on the surface of his brain, which have become stretched as his dehydrated brain tissue has shrunk. The brain itself is insensate, with no pain receptors of its own, so it is only the pain receptors in the walls of these mechanically disrupted blood vessels which can convey the signs of headache. These, that is, and the changes in various neurotransmitters such as prostaglandins, histamine and serotonin.

Some of Adam's symptoms are directly attributable to the fact that he hasn't had a drink for several hours and not to the alcohol itself. These symptoms have prompted one of Adam's harder-drinking friends to claim that the best way to prevent a hangover is simply to keep drinking. But Adam knows that isn't a sensible option; it's a sure way to alcoholism. So when he remembers he opts instead for a pint or two of water before bed to rehydrate.

Fortunately, Adam doesn't suffer from alcohol withdrawal syndrome, which usually only occurs in people who drink regularly to excess. The central nervous system compensates for the continued presence of high levels of alcohol by upgrading or downgrading various neurotransmitters to counterbalance its sedative effects. Fundamentally, alcohol is a nervous system depressant, a paradoxical concept to many people's minds since drinking usually makes them socially more confident, physically more active and generally dis-inhibited. It is the compensatory reaction of the brain to this depressant effect that results in the tremor, sweating and palpitations witnessed in chronic drinkers. The depressant effect of alcohol can be better understood perhaps by acknowledging what aspects of brain activity are reduced. It is the very system of social interaction which regulates and moderates our behaviour in relation to other people that becomes depressed. We therefore become less than usually self-critical and controlled. We lose our protective inhibition. Our natural inhibitory mechanisms are depressed. Adam is a social drinker rather than a heavy regular drinker, and tends to drink only occasionally when in company rather than on a daily basis even when alone, yet all the evidence suggests that his hangover is in fact a mild manifestation of the same physiological changes that occur with alcohol withdrawal syndrome in drinkers who are heavily alcohol-dependent. So what about all the toxins and other chemicals coursing through Adam's bloodstream as a result of last night's bingeing?

When alcohol is metabolized and broken down by the body, it involves a two-step process. First, an enzyme called alcohol dehydrogenase changes alcohol into acetaldehyde and then a second enzyme changes that into acetate. Higher concentrations of acetaldehyde cause unpleasant symptoms such as fast pulse, skin flushing, nausea and sweating, but usually it is broken down efficiently and quickly. Some people, however, are genetically less able to do this and feel unwell after drinking relatively modest amounts of alcohol.

Alcoholic drinks don't just contain ethanol of course. Other biologically active compounds called congeners can be produced during fermentation and these influence the taste, smell and look of the drinks. The hangover effects of congeners should not be underestimated. Drinks containing more pure ethanol such as vodka and gin tend to produce fewer hangover symptoms than drinks containing large amounts of congeners such as red wine, brandy and whisky. Methanol is one of the congeners most implicated in the cause of hangovers, as it is particularly toxic when it has been metabolized into formaldehyde and formic acid. After all, formaldehyde is what dead human bodies are pickled in after their previous owners have donated them to medical science for dissection.

Adam had heard that some people maintain that drinking red wine causes them problems, but not white. Why would that be? Congeners aside, recent research has demonstrated an increase in levels of plasma serotonin and the allergy mediator histamine after consuming even the finest clarets. It seems that antibodies to the protein in certain grapes, just like the protein in kiwi fruits or wheat, can lead to delayed hypersensitivity and unpleasant symptoms in susceptible people. Adam's problems, however, are not to do with allergy or food sensitivity or alcohol dehydrogenase deficiency. Congeners play a part, largely thanks to the several whisky chasers he'd rashly downed the night before, but the predominant cause of his hangover is simply the volume of ethanol he consumed. It is completely self-inflicted, as he knows all too well.

Hopefully, he is now thinking, his self-administered heavily researched hangover cure will abolish his symptoms soon. It needs to, because Adam's date with destiny, and his eagerly anticipated sex lecture, is only hours away.

Chapter 27

THE BELL HAS sounded to start the school day at Ben's college and as he trudges towards the main entrance hall he notices how worn down Paul's shoes are on the inside of the heel. He glances across at Tom's but they're not the same. His are scuffed, dirty and splayed at the front with a tiny hole just beginning to appear above his big toenail, but the heels are flat and symmetrical. Ben's interested in feet because his often ache and the verrucas he has got on the soles of both are even more likely to make him feel like he's treading on tiny pebbles when he's running in an inter-collegiate sports meeting this afternoon, competing in the mile.

He's a good runner, Ben, with a natural talent for middle distance. He's got his aim firmly set on retaining the coveted Wilmslow Cup which he won last year in record time. But his feet, he realizes now, will hurt afterwards. Just thinking about the event immediately sends a tiny surge of adrenaline through him. He feels it physically in his stomach and chest. A few hundred molecules of noradrenaline released from the adrenal glands above his kidneys rush upwards in his blood-stream to his intestine, heart and brain. There is a physical kick he feels in his midriff, the flutter of butterflies some-where under his ribcage and all the hairs on his body stand up on end. Momentarily he feels a thumping in his heart and a distinct flushing of his skin and then it's gone. He recog-nizes this pre-race anxiety and he can deal with it. It's normal and he knows he can use it to his advantage. But it's too early just yet, so he turns his attention back to Paul's shoes. Paul is obviously a pronator. When he walks or runs, his foot rotates outwards on his ankle joint and his heel strike is on the inside of his heel, hence the worn-down shoe. Pronation can occur very early in someone's development. It can also be the result of foot strain from excessive weight on the arch or from weakened ligaments in the same area in old age. It can even occur as a result of wearing high heels for long periods but this, in Paul's case at any rate, Ben feels is unlikely. Paul may be a bit of an old woman at times, he thinks, but on the balance of probabilities he isn't someone Ben can imagine as a cross-dresser. Tom, maybe. But Paul, no.

Pronator, yes. Female impersonator, no. Podiatrists and the experts at the running shop where Ben gets his trainers would advise Paul to buy shoes to compensate for the pronation, or to wear Sorbothane inserts to avoid problems with his feet, knees, hips and back in later life. But then they would also agree that probably three-quarters of people wear the wrong size shoes anyway. Most people continue to wear the same size of shoe they were measured for years previously, not realizing that, as time passes by, their foot size and shape can change considerably. In fact, the size of our feet changes significantly at different times of every day. When you stand on them, they enlarge and spread and after a few hours in that upright position they swell a little as a result of the normal gravitational fluid leakage from the blood capillaries. That's why shopping for shoes is always best done in the late afternoon. Any time is a good time as far as most women are concerned, but the best time in physiological terms, at any rate, is the evening.

None of us is symmetrical either. Almost all of us have one foot which is larger or wider than the other. Consequently, we should all purchase shoes which fit the larger foot, since shoes are only sold in matched sizes and buying shoes which are too small may be fashionable but is likely to result in crippling pain. Feet, on the whole, are such a neglected part of the human body. Furthest from our minds in both the physical and caring sense, they are nevertheless vital to our mobility and independence and highly indicative of our general state of health. It is our feet which often show the earliest signs of arthritis, including the systemic metabolic disorder known as gout. It is our feet which give away tell-tale clues about the presence of diabetes or poor circulation. One-third of all the bones in the human body are located in our feet, with 26 of them in each one, together with 33 joints, 19 muscles, 10 tendons and 107 ligaments.

Of all the tendons in the foot, the Achilles tendon is unquestionably the most important. Named after the

Ancient Greek hero of the Trojan war, this tough fibrous band of tissue is the thickest and strongest in the body. According to legend, Achilles was invulnerable except for this one area at the back of his heel. In Ovid's 'Little Iliad', Achilles' mother Thetis attempted to immortalize her son by dipping him into the river Styx, holding him only by his heel but in doing so making it the only part of his body susceptible to a mortal wound. The tendon connects the gastrocnemius muscle at the back of the lower leg to the heel, allowing the forefoot to push strongly downwards when the calf muscle contracts. It allows standing, walking, running and jumping, each Achilles tendon withstanding the entire weight of a person's body with each step. During the push-off in a sprint, that load can increase by as much as 3–12 times. Not surprisingly, an Achilles tendon rupture is a frequent and incapacitating injury, common in middle-distance runners like Ben, as well as sprinters like Usain Bolt and racquet sports players. The tendon can suddenly snap with the sound of a gunshot, leaving a foot incapable of pushing up off the ground and the calf muscle bunched like a clenched fist, abnormally high up behind the back of the knee.

Achilles tendonitis is bad enough. Ben had suffered this for a few weeks himself last year. His tendon was stiff and painful when he got out of bed in the mornings and creaked like the unoiled hinges of a door whenever he walked. It had taken two months of painful deep transverse massage from the physio-terrorist, as Ben called her, and reluctant but enforced rest to put it right. It doesn't have much of a blood supply, she had carefully explained, so it heals slowly, and the calcification that forms within it takes a long time and considerable mechanical manipulation to break down.

Achilles tendon problems, like so many others affecting the feet, often come about through negligence. Women who spend fortunes on fashionable shoes, and who love nothing more than to cram their feet into killer-heel stilettos with pinched pointed toes, neglect their feet in hundreds of ways. But at least while they are wearing high heels, their Achilles tendons are to a large degree safe because of the shoes' action to lift the heel and relax the tendon. Not so, however, for the bases of their big toes, which are highly susceptible to bunions, and their other toes, which frequently exhibit the phenomenon of 'overriding' where one slips under or over its neighbour. Calluses and corns are common too, together with ingrowing toenails, fallen arches and fungal infections. But what are these various conditions? Why are they so common and what causes them? Why are our feet so smelly too? And is it true, as Ben is playfully suggesting to his friends right now, that the boys with the biggest feet also have the biggest penises? Tom, whose shoes are a size 12, definitely thinks so. Joel, in a narrow-fitting size 7 shoe, has just gone strangely quiet.

The average person takes about 10,000 steps a day, and in a lifetime probably walks about 160,000 kilometres. That's the equivalent of walking around the earth four times. We normally start toddling around the age of one, some managing it a little earlier and some a few months later. But the feet of a newborn baby are disproportionately long. Not only that, but in the first year of a child's life their feet grow rapidly, reaching almost half their adult size in those initial 12 months. By the age of 12, their feet will be about 90 per cent of their adult length. In time, they will support around 250,000 sweat glands, together capable of secreting some 500ml of perspiration daily. And the skin under the sole of the foot will become 20 times thicker than the skin anywhere else. This moist expanse of thick layers of dead skin cells is manna from heaven as far as fungal micro-organisms are concerned. Once they and the millions of bacteria which live on our feet get to work on the sweat and break it down into aromatic amine compounds resembling those of overripe cheese, the odour given off can be horrendous.

In evolutionary terms, humans are bipeds and planti-grades. That is, we walk on two feet and use the whole undersurface of our feet, not just our toes like dogs, cats and birds do. In this respect, we are similar to primates, bears and frogs. Millions of years ago, we would walk barefoot with just primitive animal skins as coverage for warmth. In many ways, our feet were all the healthier for it. Research com-paring the feet of 2,000-year-old skeletons to their contemporary counterparts demonstrates much healthier feet then than now. And Zulus, who still go about largely unshod, have the healthiest feet on the planet.

On the subject of skeletons, the correlation between foot size and pelvis size certainly holds true in obstetrics. A pregnant woman like Eve with relatively large size 7 feet is genuinely more likely to have a wider pelvis and therefore a birth canal more favourably disposed to easy childbirth. But

Tom would be disappointed to learn that any similar correlation between shoe size and penis size is a myth. Even more disappointing to learn that a survey of 104 men, carried out in 2002 by nurses at St Mary's Hospital and University College Hospital in London, actually disproved this theory. Common mechanical problems of the feet unfortunately are extremely real. And horribly inconvenient and painful. As the Ancient Greek philosopher Socrates famously once commented, 'When our feet hurt, we hurt all over.'

Eve's mother, Grace, would agree with him. She suffers constant pain in her feet because of bunions. A bunion is a bony deformity on the base of the big toe otherwise known as the metatarsophalangeal joint. It produces a lump here when the big toe starts to migrate away from the midline of the body towards the second toe, eventually leading to discomfort and pain. The overlying skin becomes red, blistered and infected and a fluid-filled sac called a bursa often forms below it, making it puffy. The actual deformity is known as hallux valgus and to some extent the tendency for this to happen is hereditary. Women suffer from it more than men, and part of the problem is the narrow high-heeled shoes they like to wear. However, shoes alone don't cause bunions, they merely make a pre-existing problem worse. Once you've developed bunions, the therapeutic options are not terribly attractive. You can wear wide flat shoes with laces or adjustable straps which look like surgical boots, or you can apply a bunion pad over the prominent bone to lessen the pressure from your shoes. Since painkillers only partially ease the discomfort, the only way to deal with the problem permanently is to submit to surgery. An exostectomy shaves off the portion of bone which sticks out, but doesn't correct the underlying positional problem of the toe. Only a metatarsal osteotomy does this. And it's painful. The surgeon cuts one or more bones in the toe, and realigns them in a straightened position using a sturdy wire which is removed a number of weeks later once the bones have healed.

Ben's grandmother, Grace, has declined the offer of foot surgery for her bunions but at least she has had a lot of sympathy from her grandson, who has foot pain of his own. Ben's foot problems are of a much more masculine nature and of a different type altogether.

Aside from his painful verrucas, Ben has an ingrowing toenail on the left side, and tinea pedis, otherwise known as athlete's foot. These are usually but not exclusively a male problem because generally men are less fastidious about looking after their feet, are less fussy about their hygiene, sweat more, and tend to wear shoes which either push down hard on the top of the big toes or are made of synthetic materials which don't allow the skin of the feet to breathe.

But why do we need toenails and fingernails at all? How fast do they grow, and what can they tell us, if anything, about our general health? Nails are basically the human version of claws. Made up of dead cells which grow from a living base or matrix, they have three main functions. First and foremost they protect the soft vulnerable pulp of the fingertip or toe tip from injury, while at the same time they increase the sensitivity of the tip by providing a firm counterpressure to the nerve receptors under the skin below. Fingertips also allow for an extended precision grip. The ability to oppose forefinger and thumb helps us pick up and manipulate small objects, but the extended precision grip of our nails allows us to perform even more intricate tasks such as the removal of splinters.

The speed with which nails grow is largely determined by the length of the terminal phalanx, the bone in the very last part of the finger or toe. That is why the nail on the index finger grows slightly quicker than the nail on the little finger. The average growth rate of the nails is about 3mm per month but the fingernails grow four times faster than the toenails. Consequently, it will take a fingernail about 3–6 months to regrow completely, but 12–18 months for a toenail to do the same. In individuals, the rate of growth changes further according to age, sex, exercise levels, nutrition and genetic make-up. Nails will also grow faster during the summer months, because the circulation in the digits is better in warmer weather. Many people believe that the nails continue to grow even after death. They might appear to, but in fact the appearance is an optical illusion. After death the body dehydrates and the skin surrounding the nails shrinks and retracts just like it does everywhere else, so the nails look as if they've grown but in fact haven't changed at all.

Ben is fed up with the pain from his ingrowing toenail and knows that running in the mile race will not help. But nails themselves are insensate. The tough protein called keratin in the dead cells which make up the nail has no nerve supply. The pain which Ben is feeling is all coming from the soft delicate skin at the edge of his big toenail into which the hard corner of his nail is pushing. Here, and below the nail in the nail-bed matrix, there is no shortage of sensitive nerve endings together with blood and lymphatic vessels. The matrix is protected by the nail, and it is from here that new nail plate cells are formed, gradually pushing the older dead ones forward so that they become compressed, flat and translucent. This translucency along with other features of the nails enables doctors to gain insight into a person's general state of health.

If the nails look blue, that is because there is cyanosis of the tissues below them, a lack of oxygen in the blood, sometimes caused by poor circulation in the fingers themselves but often seen in people with chronic lung disease such as chronic obstructive airways disease or lung cancer. It is a common feature of heart disease as well. If they suspect this, doctors will also look closely for signs of 'clubbing' – not evidence of wild partying into the early hours, but the development of patulous (broadened-out) fingertips with widened club-shaped fingernails to match. It's another indication of serious underlying disease of the lungs or heart, or of chronic inflammation of the bowel. In fact the nails can exhibit a variety of different colours. Green if the nail is infected and has separated and lifted away from the nail bed. Yellow and crumbly in the presence of fungal infection. Brown in heavy smokers due to nicotine staining. Dappled white when there has been previous minor trauma or a nutritional deficiency such as lack of calcium.

The shape and texture can be revealing also. Longitudinal ridges running from the lunula (the paler moon-shaped crescent at the base of the nail) to the tip may just be a typical sign of advancing years. So might onychogryphosis, where the nails become thickened, distorted and horn-like. However, ridges also develop in people with rheumatoid arthritis, eczema or another chronic skin condition known as lichen planus. Tiny widespread pits in the nails are closely associated with psoriasis. Concave, spoon-shaped nails otherwise known as koilonychia are seen in chronic iron deficiency anaemia.

Among Ben's friends, and in adolescence in general, it is estimated that as many as 44 per cent will show evidence of nail biting – onychophagia in other words. A stereotypic movement disorder, this is on a par with hair playing, thumb sucking, fist clenching and compulsive scratching and is frequently just a physical manifestation of stress. Considering all the damage we inflict on our nails either deliberately or accidentally, and all these medical conditions which can take their toll on them, it's hardly surprising that nails often appear visibly unattractive and ropey. Women particularly know just how noticeable nails are and understandably like to make the most of them. They trim them, shape them and smooth them. They rub them down with emery sticks, push back the cuticles and paint them. Nail salons offering nothing but nail care have sprung up and are flourishing, and the market for false nails and nail art has never been greater, tripling in size in the last 20 years. Little wonder that the manicure industry as a whole is worth a staggering $20 billion in the US alone.

Unlike doctors, who are more concerned with one of the most common conditions on the planet, foot fungus or tinea pedis, beauticians are more concerned with nail beauty. Using up to 10,000 different chemicals in their products (about 90 per cent of which have never been safety-tested by any independent agency), they can certainly make nails look interesting and attractive. Yet there are concerns that un-regulated nail products could have the potential to cause harm. Contrary to popular belief, our nails are not as impervious to the outside world as we would like to think. Our nails can absorb certain substances, including chemicals like the garden weedkiller paraquat, which is potentially lethal. The pursuit of beauty, however, usually takes precedence over any minor worries about health and some folk have even taken matters to extremes. According to the *Guinness Book of Records*, in 1998, Shridhar Chillal of India had nails on his left hand measuring 615cm in total length, and the thumb nail extending to 1.5 metres in length on its own. His female US counterpart, Lee Redmond, in 2008 had nails on both hands measuring a total length of 8.5 metres.

Ben, hobbling slightly into his first class of the morning and trying to forget about the athletics scheduled for later, would happily go without a big toenail at all. It isn't life-threatening, he realizes, but it *is* tender. He's tried soaking it in warm salty water to soften it and then pushing a tiny pellet of cotton wool under the side of the nail to lift it away from the flesh, but it just isn't helping. Yet he doesn't really fancy the doctor's other suggestion: anaesthetize the toe, cut along one-third of it longitudinally and then peel it back to the nail bed with surgical forceps, like lids are peeled off the tops of sardine cans. That sounded like torture. It would be done under local anaesthetic so he shouldn't feel anything, but it isn't something Ben is keen to contemplate. No. Ben has decided that if he is going to do anything to his feet he will get those tiresome verrucas sorted out first. They are warts after all. And Ben doesn't like the idea of warts growing anywhere on his body.

Caused by the human papilloma virus, verrucas are the same as warts except that they are situated on the sole of the foot. That's why they are medically known as plantar warts. It is precisely because of their location that they can become very painful. Burying themselves deep into the thick dead outer layer of skin under the feet, verrucas becomes rock-hard and pebble-like. It can feel as if you are walking on a marble. There is often a visible black dot in the middle of a verruca, which is just a small blood vessel seen end-on, and there is normally a pale flattened area around it. From here, because the outer layer of skin is constantly being shed like the redundant skin of a metamorphozing snake, the virus can spread to other areas of the feet or, of course, to other people. The human papilloma virus is ubiquitous in nature, and while it is certainly not highly contagious, it does have the capacity to be passed on to others if they have the tiniest cut or roughened area in any part of their skin, and are in direct contact with it. It is also much more likely to be picked up by people whose immune systems are compromised and weaker for any reason.

Ben knows, from his surreptitious observation in the boys' changing rooms, that at least three other pupils in the athletics squad have got verrucas. That is consistent with the recognized figure of 10 per cent of the population having them at any one time. One boy has flat verrucas like Ben and is using salicylic acid plasters on them. Another has filiform verrucas, so called because they are elongated and fingerlike, and this boy regularly applies glutaraldehyde solution to them and then rubs them down in the mornings with a pumice stone. Ben's been sticking a simple bit of duct tape on his verrucas. He had read that this approach is just as effective as these messy chemical treatments and, more importantly, it is very cheap. The same doctor who had offered to mutilate his toenail had then generously suggested the alternative approach of the application of liquid nitrogen. The more Ben thought about it, the more he thought that his doctor was beginning to sound like a sadist. A bit of a budding Harold Shipman. In other words, he thought the treatment was well worth avoiding.

In a way, Ben was right. Not about the doctor, who was only offering the best treatment available, but about avoiding treatment altogether. The wart virus is simply an infective organism which sooner or later sensitizes the immune system to respond. Consequently, three out of ten verrucas will disappear spontaneously without treatment within ten weeks and almost 100 per cent will have cleared up within one to two years. Warts elsewhere can take a little longer, as without the weight of the body pushing them into the underlying skin and traumatizing it, the immune response is more sluggish. The trouble is, if they're painful or rapidly spreading in the early stages, treatment is sometimes required. Salicylic acid treatments are supposed to clear 70 per cent of them within three months, and other treatments boast similar success rates. The chemical treatment can be fiddly and messy, however, so liquid nitrogen application, which is done by GPs or in hospital outpatient clinics using a cylinder of the liquid gas, can be a shorter, sharper option. This cryotherapy technique freezes the verruca and its surrounding area and raises a blister within the next few days which kills the virus and then heals. Even then, several applications may need to be given.

Ben hasn't yet found the time or the motivation to apply anything other than duct tape. He refuses to wear special antiviral socks at the swimming pool (you can only imagine just how badly he'd be 'rinsed', as he calls it, by his friends if he did) but he has started to wear flip-flops when he showers and is using the internet to investigate a new treatment with a medicine called Imiquimod, also used for skin cancers, which stimulates the immune system to attack the virus and therefore eradicate the problem more quickly.

Lately, as a result of his background reading for his biology lessons, Ben has felt rather envious of the humble centipede. With its 100 feet, the strain must be so much less on each pair, he imagines. What he doesn't know is that centipedes never actually *have* 100 feet. Despite being extensively studied for over a century, not one has ever been discovered with exactly 100 feet. Some have more, others fewer. One, found in 1999, had 96 legs, and was unique among centipede species in that it had an even 48 pairs. Other centipedes have odd numbers of pairs of legs ranging anywhere from 15 to 191 pairs. With multiple pairs of feet, Ben thinks, centipedes can at least take the pressure off any one of them if it suddenly becomes sore. On the other hand, with only one pair of feet like him, there must be less that can go wrong in the first place. Imagine having fungal nail infection and having to apply pharmaceutical paint for up to a year and a half to around 100 feet. Or having to swallow so much of the antifungal terbinafine orally that you would get liver toxicity as a result. No. On balance, he concludes, two feet are probably better.

Nevertheless, with previous Achilles tendonitis, fungal infection, ingrowing toenail and verrucas, Ben is of the opinion that his feet aren't exactly his strong point. But he's wrong. Ben's feet will later be powering him around the athletics track and going for another college record. And all his current afflictions are on the *outside* of his feet. They are not causing any internal, mechanical disruption and are eminently treatable.

Chapter 28

BACK AT HER own school down the road, Poppy is hiding out in the toilets to attend to a huge acne spot that has actually got bigger since she first noticed it in the mirror this morning. She would have dealt with it there and then had she not been distracted by her brother's unwelcome threats to intrude into her bedroom.

Why does Ben always have to come into her room? He knows she hates it, but that's probably why he does it. Maybe she should try a bit of reverse psychology and invite him in as often as possible. Chances are he'd do the complete opposite and question why on earth he would want to come into his sister's room anyway. As far as Poppy is concerned it is a major irritation. He never ever knocks and he'd hate it if she ever went into his room. Not that she'd want to. Why would she want to contaminate herself in some smelly boy's room?

'My older brother's disgusting,' she says to herself. 'And so are his friends. Except maybe Frankie. Frankie's OK.' She wouldn't want him to see her with this Vesuvius of an acne spot though, would she? And it is monstrous. Right on the end of her chin like a beacon for all to see with lots of imaginary arrows pointing at it, screaming, 'Take a look at this monstrosity, why don't you?' Why does she have to get spots at all? she asks herself. Couldn't doctors have found a way by now to prevent them forming, or to block the hormones that cause them in the first place? They say they know the underlying reasons why acne develops but seem totally incapable of coming up with a treatment that works.

She has tried creams, lotions, facial scrubs and gels. She has tried tablets and capsules and devices which shine blue light and ultraviolet light and all sorts of other lights in between. Her doctor gave her the oral contraceptive pill Dianette, in the hope that the cyproterone acetate it contains as well as the contraceptive would block the effect of the hormones which are causing the problem. Better not let her mum find the packets though, she's thinking, because she hasn't told her she's taking the pill, even though she doesn't need it for *that* reason, and it would just be a hassle explaining it.

Puberty is such a drag. She is glad it's happened, and is still happening, but it never seems to work out in her favour. She doesn't know what to expect, and she doesn't have any control over it. Why has her friend Angie got nicer boobs than her already? Why does Lyndsey suddenly look so sexy and get all that attention from the boys? How come none of her friends gets a massive great pus-filled mountain of a spot like this one? It just isn't fair and quite rightly she wants to know why nature's picking on her, and making her pick at her spots for that matter.

The fact is, she can't expect all those attractive things about puberty without suffering some of its disadvantages as well. She is producing more of the sex hormone oestrogen, which is giving her curves in all the right places and making her feel sexier, but it's also responsible for that increasingly frequent low tummy pain and rather unpredictable periods, and, even if she doesn't want to admit it, some fairly traumatic mood swings. She is also producing a greater amount of testosterone. This would have surprised her as it's the male sex hormone, which she'd always imagined only people like Mike Tyson, Bruce Willis or Superman had in abundance. Actually, just like other girls her age, she's producing it too, albeit in much smaller amounts than boys going through puberty, and that's what's responsible for the growth of her body hair in all those private places, her stronger muscles and an increasingly assertive attitude her family are beginning to see more often. As if! Poppy? With attitude?

Maybe all this stress is making her spots worse, she is thinking. Or is it the occasional bar of chocolate or small portion of French fries at McDonald's? What if it gets worse and ends up leaving her with scars on her face like it did to her friend Mary's older sister? She's still getting spots into her thirties. What if Poppy's never get better? What if she constantly looks like a pepperoni pizza or something hideous out of a dermatology horror movie? And should she actually squeeze this spot and evacuate its disgusting contents, or leave well alone as the doctor told her to do? These questions may well represent the most important focus of her entire day, maybe her entire week. Because it'll probably take that long for this spot crater to heal. Even wearing make-up won't hide the damn thing. Her doctor had told her to avoid make-up at all costs, for fear it will just clog up the pores even more.

Acne is caused by an increase in the production of a thick oily secretion called sebum, made by the sebaceous glands in the skin. Normally, this keeps the skin supple, soft and stretchy, and lubricates the hair follicles. During puberty however, when testosterone production increases and the glands become more responsive to its action, the oil becomes thicker, turning from a thin sun-lotion kind of consistency into a gloopier, peanut-butter kind of goo that can block the pores and expand sideways into the surrounding skin. And if it can't come out on the skin's surface, it will just keep on building and push out sideways causing the spot. If it stays that way, it looks pimply and white. A whitehead. But the moment the air gets to it and bacteria invade, the top turns dark and it becomes a blackhead. Poppy might not like her spots, but the bacteria, by contrast, love them. As the sebum changes into chemicals called free fatty acids, the germs are attracted to the spots like bees around a honey pot. With this infection comes white blood cells which make pus, and dilation of blood vessels around the spot causing redness.

Of course she can squeeze the spot if she wants to. If it's ripe and very near the surface. She can pop it by using two fingernails either side of it and pressing hard from the bottom up. But she really shouldn't do it if it's a young spot. A baby spot, deep down in the gland with a thick lid of dead skin cells over the top with nowhere to go but sideways. The result would be hugely disappointing and only inflame the surrounding skin even more and make the redness much worse. What she really needs to do in this situation is to apply a flannel soaked in water as hot as she can bear, and keep it pressed against the spot for a minute or two before repeating it. This increases the circulation around the spot and encourages healing from within, or at least speeds up the process. On the other hand, if it's ready to be popped, she should get on with it now. Once burst, an antibacterial lotion is useful to cleanse it and to prevent the spread of germs to other pores.

As for her diet, Poppy really doesn't need to avoid the occasional treat of chocolate or chips. There is no truth in the idea that eating oily food leads to oily skin because the oil Poppy consumes is completely altered by her digestive system long before it reaches her bloodstream and even longer and more effectively before it gets a sniff of her skin. The Dianette contraceptive pill the doctor gave her will help because the cyproterone it contains will block the effect of testosterone on sebaceous glands. Benzoyl peroxide preparation from the chemist will exfoliate some of the dead skin cells on the surface of her skin which contribute to the blocking of the pores, and failing that, an antibiotic lotion such as Clindamycin will keep the bacteria under control, as will an antibiotic like tetracycline or erythromycin taken orally at low dosage for several months. As a last resort,

Isotretinoin creams such as Retin-A from her doctor are another useful option, as they chemically shave off and thin out the surface skin cells. Poppy can always try taking several of these in combination.

She did once ask her doctor if she could be referred to the skin specialist at the local hospital and have a course of Isotretinoin capsules like her friend's older sister. While the doctor thought that Isotretinoin was extremely effective, he felt her acne wasn't anywhere near severe enough to warrant the possible side effects such as blood disorders, joint pains and suicidal imaginings. He had pointed out to Poppy that Isotretinoin could even damage unborn babies. Idiot, Poppy had thought at the time, shows how much *he* knows. She is unlikely to be pregnant as she isn't sleeping with anyone, and suicide can almost seem a better option than being at school with this monster of a spot on her face.

Chapter 29

Poppy's morning has dragged on today through a French lesson, a biology tutorial and double maths. Her lunch break could not have come soon enough. As she contemplates the limited menu she cannot help wondering why some people become fat and others don't. What keeps some of the girls in Poppy's school stick-thin, whereas others seem to pile on weight inexorably until they can hardly move?

Poppy is in the queue for cold foods and salads. Sinead, the same age as Poppy and already a size 16, weighs 14 stone and is queuing for pizza. How much of the UK's current obesity epidemic can be explained by too many people making the wrong food choices? Poppy wonders. We are what we eat, they say, but how come some of the girls seem to scoff voraciously all the time and look like whippets, whereas others eat like birds and look like Shrek? Are they just genetically programmed to be that way? Or is it all down to the levels of exercise we take and the fact that sitting around watching TV all day doesn't burn off any calories?

Poppy is relatively self-conscious about her weight. She keeps a close eye on her figure in the mirror every morning, checking to see if her clothes are getting tighter in any of the wrong places, and weighs herself at least three times a week. She reads the nutritional labelling on the food Eve buys and keeps a rough mental note of her daily calorie intake. Unlike her brother Ben who loves exercise and has a particular talent for running, she exercises regularly but only as a means of keeping her weight in control and burning off some of the calories from those naughty energy-dense foods she sometimes can't resist such as M&Ms, mint Aeros and Krispy Kreme doughnuts. The one form of exercise she loves and excels at is pole dancing in the gym, which, when performed in the way she does it, involves grace, elegance, strength and impressive gymnastic ability. This very afternoon she will be perfecting her spins, inverts and climbs. She will also be trying out a couple of new moves called the Crucifix and the Butterfly.

Poppy is body-conscious and careful about what she eats, but she doesn't have an eating disorder and unlike at least three girls in the school she isn't anorexic. The three in question are all very similar, although they are by no means best friends nor are they in the same classes or year groups. They are conspicuously thin with pale, almost translucent skin and particularly prominent bones around their clavicles, shoulder blades, pelvic girdle, knees and hips. They all have fine downy hair on their forearms and they all dress in loose-fitting clothes, thinking nobody will notice how much weight they have lost. They hate it if anyone calls attention to their diet or shape and always seem to have an excuse for not sitting down and eating with everyone else. They will say they've already eaten. Or that they had a huge breakfast. Or that they are feeling a little sick today. Cherie, who has been in and out of hospital a few times already and is off school at the moment, has never been seen eating by anybody, Poppy thinks.

Anorexia nervosa, once a rare condition, is increasingly common in schools like Poppy's and most of the girls, if not the parents who are of a generation when anorexia was almost unheard of, are painfully aware of it. They know that sufferers are constantly in denial about their diet, and are secretive about the tricks they use to avoid eating, to dodge foods which are higher in calories, and to decline eating in company. They have heard about their obsessional fear of being fat, their tendency to abuse laxatives, their self-induced vomiting and their preoccupation with prolonged vigorous exercise as a way of shedding weight. Simone, who looks like a ghost and jogs every day, could probably even give Ben a run for his money in a foot race. Anorexia nervosa stalks Poppy's school like the grim reaper on a scouting expedition. The girls are all aware of the victims in their midst, and of the strange psychological changes they are going through.

But in some of the other girls at least, there is also an in-explicable feeling of envy, a desire to emulate or even compete with them, in the pursuit of what so many girls, magazines, newspapers and TV programmes are insisting is the only desirable type of female body. A supermodel's body, a size zero, a wafer-thin frame with prominent cheekbones and sunken eyes, total absence of any muscle definition. It is a look. A fashion. But it is also a goal anorexics seek above all else in that morbid fear of fatness and in that desperate attempt to discover perfection, to avoid the challenges of growing up out of childhood into womanhood with curves and healthy desires and to exert vice-like control over a changing world around them. The spectre of anorexia is constantly there at the school, and the teachers have delivered many a tutorial on the subject and are always on the lookout for any pupil showing overt signs of succumbing. But what chance do they have of doing anything constructive about it? In a condition which is so difficult to treat or to alter the course of, even in parental or professional hands, a schoolteacher's task is an uphill one to say the least.

Sinead, on the other hand, is at the other end of the eating disorders spectrum. Obese, even morbidly obese. Why is it, Poppy asks herself, that she feels saddened and sorry for the girls with anorexia nervosa, but critical and censorial about the overweight girls and boys like Sinead? Even the terminology she uses tells the story.

With Cherie, she wants to throw her arms around her and tell her everything will be all right. She wants to help her with any problems and wrap her up in comforting cotton wool. With Sinead, on the other hand, she just wants to insult her. Poppy knows it is illogical. She knows that it is cruel and hurtful to tease people about being overweight, and to blame them for their condition. She even knows that it is precisely teasing that can lead to the psychological damage that triggers anorexia nervosa, the trigger which originally motivated an overweight Cherie to lose a colossal 5 stone in six short months. So there is a paradox within Poppy herself, she realizes. There is an instinct to abuse and hurt Sinead, who is fat, but to love and comfort Cherie, who is thin.

Sinead stands there alone in the pizza queue, her down-turned mouth exaggerating the contours of her double chin, her chubby arms with their blotchy skin dangling by her side and the rolls of fat spilling out around her waist, back and thighs. She waddles forward slowly towards the serving counter, head down. Now, suddenly, Poppy feels sorry for her too. A pang of pity. Sinead, she realizes, probably has no self-esteem whatsoever. No self-confidence. She never speaks to anyone other than Poppy and no one ever speaks to her in return. She probably doesn't think anyone would show any interest, and maybe she genuinely doesn't believe she has anything to offer. It is a self-fulfilling prophecy which increasingly pushes her towards the escapist behaviour of comfort eating in which she is now indulging. Poppy, rather ashamed of her own thoughts, suddenly shouts across at her.

'Hey, Sinead, you OK?'

Sinead looks up rather warily as if half expecting some cutting remark to follow.

'Yeah. Yeah, fine, thanks. You?'

'Cool,' says Poppy rather guiltily, remembering some of the banter she has indulged in about Sinead behind her back. There was the famous joke one of the other girls had shared with them only yesterday. How did it go again? Jimmy Carr had been performing one of his brilliant stand-up comedy routines. This one had majored for ages on people who were fat or obese. After the show one very overweight lady had gone backstage to berate him. Barging into his dressing room unannounced, she confronted him head-on. She demanded an end to this disgraceful, politically incorrect lampooning of people her size.

'You, sir, are fattist!' she declared, prodding him with one plump forefinger.

'No,' he had calmly replied. 'Compared to me, if you have a careful look at yourself, I think you'll find *you* are fattest,' after which the fat lady had retreated to the door in floods of tears, murmuring, 'How do you think we feel?'

'All squidgy, I imagine,' said Carr with his usual straight face. Or so the story went. It was wholly apocryphal.

Poppy finds herself considering the issues here. Is she 'fattist', she asks herself? Does she discriminate against fat people in her own life? Does she have fat people as friends? Would she go on holiday or have a night out with them? Should she even be calling them 'fat'?

Should fattism be a hate crime as some people have recently suggested? she wonders. Should discrimination against fat people be criminally outlawed along with discrimination based on race, gender, age, sexual orientation or disability? Certain lobbying parliamentarians have already raised the question in the House of Commons. In her heart Poppy doesn't believe that you could ever really legislate against the way people feel and against any discriminatory judgements they might make. Nor does she think bringing in specific laws would change things or be enforceable. She does, however, believe that drastic action needs to be taken to stem the burgeoning epidemic of obesity that is gripping the country in which she lives. Thousands of others, including the vast majority of health care professionals, agree with her.

Recent statistics show that in the UK at least 44 per cent of men and 33 per cent of women are overweight and on top of that, one-quarter of all adults are actually obese. The distinction lies in their body mass index, or BMI, a calculation based on their height and weight. Poppy's BMI is her weight in kilos divided by the square of her height in metres. She weighs 60 kilos and she is 1.65 metres tall. Therefore, her height multiplied by itself is 1.65 x 1.65 = 2.72 and when her 60 kilos are divided by 2.72, it gives her a BMI of 22 kilos per metre squared. Poppy's BMI of 22 falls within the healthy range of 20–25. Overweight people are classified as having BMIs between 25 and 30 and obese people above 30. Morbid obesity is defined as a BMI of 40 or higher. If the current statistics are not bad enough, with at least one in three children between the ages of two and 15 already overweight, the outlook for the future is even more dire. If trends continue, half of all men in the UK will be clinically obese by the year 2030.

'Oh my God,' thought Poppy, when she had first heard that. 'That's only 17 years from now. There'll be hardly a single fit guy left to date.' What she didn't know was that women themselves wouldn't be far behind. By then four out of ten of them would be similarly obese.

Poppy had seen some of the latest newspaper headlines and was horrified. She had read the story involving 17 fire-fighters and paramedics who'd battled for eight hours to free a 40-stone man after he had fallen ill at home. And about a 19-year-old girl who at 63 stone had had to have the internal walls of her house knocked down and windows removed so that an adapted crane could lift her into an ambulance. Under the watchful eyes of doctors and nurses, social workers and policemen, she was then taken to hospital for the urgent treatment her gross obesity necessitated. And as far as Poppy was concerned, it was 'gross'. At least 200 Britons, she had discovered in discussion with her friends, were now unable to leave their homes because of their enormous size. Costing the NHS £16 million a year already, obesity was a drain on the taxpayer the nation could clearly no longer afford.

But how do people ever get to that size in the first place? she wonders. Most people, when they realize they are out-growing their clothes for the second or third time, rein in their weight gain. Their breathlessness on exertion and difficulty in reaching down to pull on their socks or wriggle into their car cannot be ignored any longer. So what allows these other individuals to cut themselves off from society and eat themselves to a premature death? And if they cannot physically walk, wipe their own bottoms or leave the house, who is buying their food and bringing it to them? What warped sense of love and devotion allows that to happen?

For all of us, weight control is and should be part of a healthy lifestyle. Obesity is inextricably linked to an increased risk of coronary heart disease, stroke and cancer. Three of the biggest killers in the Western world. Thirty thousand people die prematurely every year from obesity-related conditions. High blood pressure, raised cholesterol, digestive disorders, arthritis, snoring and sleep apnoea all take their toll. Then there are the complications which may not be life-threatening but which are distressing and life-changing all the same. Infertility. Social isolation and depression. Being the object of silent or overt insults. So why are so many millions of us ignoring the warnings? What is the underlying cause?

The easy answer is that we are consuming far too many more calories than we are burning off. It is a simple equation. Calories in should equal calories used. That means our weight will not change. But when calories in are in excess of calories burned, that means weight gain. And every day that happens, the weight gain increases. In theory, it all comes down to this fundamental balance, but additional factors such as genetics, socio-economic and psychological issues, illness, exercise and limited food choices play an important role.

Poppy, like everyone else, needs a certain amount of energy every day just to survive. She burns calories every time her heart beats, with every breath she takes, and even when she digests the food that is providing those calories. She burns calories when she sits down at her computer or watches TV and she burns calories even when she is asleep. This is her basal metabolic rate, the amount of energy she expends just to keep the cells in her body ticking over and fulfilling their basic functions when she is at rest and when no additional influences such as a recent meal or exercise are imposed. She needs 1,700 calories a day, even before she does anything else. The more active she is, the more calories her body will need but any excess calories she takes in will be stored as fat. Conversely, when she eats less than her body needs and exercises more, she forces her body to use its existing fat stores for energy. When she does this, she will lose weight.

Poppy produced an essay recently for a biology project to prove it. Housewives in the 1950s, she had researched, ate significantly more calories than their modern counterparts, but were measurably slimmer simply because their daily lives involved so much more physical activity. These days, people are many times more sedentary. We rely much more on cars instead of walking, lifts or escalators instead of stairs and TV at home rather than physical pursuits in the great outdoors. Just a little bit of regular daily exercise can make a big difference. But while as a population we are taking less and less exercise, we are also eating much more energy-dense food. High-fat, sugar-rich foods with a preponderance of starchy carbohydrates has meant our calorie intakes have increased enormously while we have hardly noticed. The size of the portions we consume has increased in parallel.

Clearer nutritional labelling has been brought in, to inform people about the food they are buying, but in truth few shoppers really understand it. The labels identify the total energy content of the food but they usually describe this in terms of kilo-calories, which represent the energy released when chemical bonds are broken during metabolism. Scientifically, by definition, one calorie is the amount of heat needed to raise the temperature of 1ml of water by 1 degree centigrade. But so what? thinks Poppy. What value does this kind of chemistry have for people like Sinead, struggling to make the right food choices? It doesn't mean much to anyone who hasn't spent years studying for a degree in nutritional science.

What it does mean is that 1 gram of carbohydrate (or protein) will produce about 4 kilo-calories of energy during metabolism, while 1 gram of fat produces about 9 kilo-calories. This demonstrates that fats are a much richer source of energy, weight for weight, than carbohydrates or protein. This makes absolute sense in physiological terms because if we were overweight by say 16 kilos due to excess fat stores, we would actually be 36 kilos overweight if we were storing the same amount of energy as carbohydrate. Fat packs more energy but weighs less. How would Sinead ever be able to shed the hundred or so pounds she needs to lose when she is too self-conscious and probably too overweight to exercise safely anyway?

Poppy works out that she could in theory achieve the weight loss if she started on a low-calorie diet of about 800 calories a day. Since a normal resting energy expenditure is 1,700 calories, Sinead would burn 900 calories a day of stored fat even at rest. Given her weight and the energy she uses up just moving it around, she could easily burn an additional 2,000 calories a day without doing any formal exercise at all, at least to start with, until she trimmed down a little and the weight loss began to slow. Then, a bit of increasingly vigorous exercise could speed up the weight loss all over again.

Poppy knows that 1lb of stored fat could provide about 3,500 calories. Sinead would have to do an awful lot of walking or jogging to lose all the weight she needed. Brisk walking, which is less economical for the human body than just ambling and would therefore burn more energy, for a 200lb person like Sinead would burn 420 calories per hour. If she did that just once a day she could increase her daily energy expenditure to 3,120 calories (1700 + 1000 + 420). Take off the 800 calories in her new diet and she could burn off 2,320 calories daily. In three days she'd burn 9,960 calories, the equivalent of 2lb of stored fat. That's 4lb in a week and over a stone in one month. In theory at least, provided she got all the essential nutrition she needed in the form of supplements, she could lose all the 8 stone she so desperately needed to lose in just eight relatively short months. Given an extra four months for occasional lapses and a natural slowing of progress, she could be the new slim person she wanted to be within a year from now.

Poppy so wished Sinead could see it this way, and had the motivation, willpower and support to do it. But she also knew that there would be plenty of people ready to sabotage her efforts. To reassure her that she was fine just the way she was now. That one more chocolate wouldn't do any harm. That it was rude to refuse a second helping. That she could postpone her diet to another day.

Could Sinead ever overcome these unhelpful influences

or would some genetic predisposition make any reasonable attempt at healthy gradual weight loss impossible? Poppy wonders. The role of genetics in obesity is questionable. Is there really a fat gene as people are increasingly saying? Do some folk genuinely have a slower metabolism? Are people justified in using medical conditions as an unavoidable reason to put on weight or are they just a convenient excuse designed to curry sympathy?

So far, scientists have discovered three kinds of fat genes. Unfortunately, from what we know about them, most people are not justified in blaming them for being overweight. The 'scavenger gene' has been talked about for some time. This gene is carried by about half of the world's population and determines how much energy from our food we use and how much we save. It is designed to hang on to every calorie we consume when food is plentiful so that we are protected from starvation in a famine. When famine doesn't happen, however, people who carry the gene quickly become obese. They also develop insulin resistance, diabetes and heart disease. Ironically, the very same mechanism designed to keep them alive condemns them to a premature death.

Another fat gene has been identified in people who are prone to a number of rare diseases. These people absorb more fat than normal people and rapidly put on weight. Another topic of excitement in the obesity gene world concerns a substance called leptin. Leptin and ghrelin are two hormones that have been recognized as having a major influence on energy balance. Leptin is a protein product of the OB gene and is produced by fat cells called adipocytes.

When higher concentrations appear in the bloodstream, it crosses the blood–brain barrier and tells part of the brain called the satiety centre in the hypothalamus that the body can stop feeding. Ghrelin, on the other hand, is a fast-acting hormone which affects another part of the hypothalamus, the feeding centre, and tells the body to start eating again. In obese people with high levels of visceral fat (fat stored largely around the abdominal organs rather than under the skin), the balance of these hormones is abnormal and leptin resistance becomes common. This means that satiety is never achieved. Hunger or the desire to eat remains constant.

The exact interplay between the various hormones (and more are being discovered all the time) is not fully understood. But the existence of ghrelin and leptin does at least give the biochemists working within the pharmaceutical industry a model on which to base the development of drugs that could one day potentially solve the global epidemic of obesity. It also gives us simple tricks right now which we can use to help us with our levels of hunger. One glass of water, for example, before every meal has been shown to reduce the amount we eat by at least 8 ounces and suppress ghrelin. Maybe Sinead should drink a couple of litres before each meal, Poppy is thinking. Anything that fills her belly without the calories.

Like millions of overweight people before her, Sinead has tried dieting, without success. She lost the weight initially but put more back on in the long run. The ghrelin story may partly explain this. Slim women can learn to lower their ghrelin level, but these levels immediately rise again with diet-induced weight loss. The more someone diets, the hungrier they get. It sounds obvious, but it is almost as if semi-starvation prompts ghrelin levels to increase. This is the basis of the theory that dieting doesn't work and that healthy well-balanced eating plans and regular exercise are much smarter alternatives.

Genetics may play a part in the causes of obesity but it can only be a small one. In evolutionary terms, the genetic make-up of the human race cannot have changed very much in the last 500,000 years. Certainly not in the last 50 years. Yet it is during this timeframe that we have witnessed an explosion in the population of overweight people. We can talk about psychological issues, comfort eating and using food as weapons of control in our lives. We can eat because we are bored, stressed or tired. We can wax lyrical about the body temperature theory, which states that we eat more in winter because we are cold, and we can point a finger at the gastrointestinal stretch receptor theory, which suggests that distension of the stomach wall, which normally signals to our brain to stop eating, fails to work in the usual way in some people. At the end of the day, however, the same old message about the energy intake and output balance still holds true. No matter what Sinead's genes, hormones and psychological issues are telling her, she needs to eat less and exercise more. If she doesn't, she may continue to gain weight and end up in a vicious circle of obesity-fuelled depression. Her risk of Type 2 diabetes is already 80 times that of someone of normal weight. If she doesn't address the issue soon she may well die before her parents and never live to see middle age.

What Poppy doesn't know, because Sinead has never told her, is that she's been to see her doctor several times already to ask for help. Poppy wouldn't think so to see her queuing for pizza rather than salad, but Sinead has already been referred to a dietician and an NHS-subsidized personal trainer at the local sports centre. Neither has proved successful. Rather reluctantly, her doctor has prescribed Sinead orlistat. Otherwise known as Xenical, this prevents up to 25 per cent of the fat in the diet from being absorbed. However, it still requires a low-fat diet in order for patients to tolerate the side effects, which include the passing of frothy motions and even the uncontrollable anal leakage of oily fluid. Consequently Sinead proved unable to stay on it. Sinead's parents are now pressurizing the doctor to arrange bariatric surgery. They feel the only answer is a gastric band or bypass. One million people are eligible for it, they argue, and some of them are having it done. So why shouldn't Sinead have it too? Surely this is the solution that would enable her to enjoy a normal life?

Gastric band surgery is certainly a highly effective method of achieving substantial and sustained weight loss. Also known as laparoscopic banding or lap banding for short, it involves fitting an inflatable band around the stomach, dividing it into two parts. It creates a smaller pouch at the top, which fills rapidly when you eat and makes you feel full. Then the food passes slowly through the opening into the lower part of the stomach and continues as normal down into the intestine. It limits the amount of food a person can eat. In addition, the gastric band can be inflated further with saline solution to reduce the size of the opening into the lower part of the stomach whenever necessary, a procedure which is reversible.

Another procedure, which is not so easy to reverse but which also reduces the amount of food you can eat and the number of calories you can absorb is called gastric bypass surgery. Here, a stapler is used to divide the stomach into two parts to create a pouch at the top, similar to the one created by lap banding surgery, but instead of the food passing down into your stomach, it bypasses the rest of the stomach and most of the intestine. Instead, a surgically refashioned and shortened length of bowel takes the food directly to a point much lower down the digestive system. This explains why fewer calories are absorbed. Rather amazingly, the whole operation can be performed using keyhole surgery, by way of five or six small cuts. Sinead's parents think that this is the Holy Grail as far as providing an answer to their daughter's problems is concerned, but unfortunately for them Sinead's GP is not exactly playing ball.

The doctor keeps reminding them, reasonably enough, that complications can ensue. He talks of lung problems and injury to the spleen. He talks of gastric band slippage and leakage at the junctions of where the bowel has been repositioned. He reminds them that bleeding, infection and blood clots may occur, as they can after any operation. He also points out the problem of excess saggy skin after significant weight loss and the fact that any psychological issues producing addictive behaviour other than food addiction will not have been addressed. He even tells them gravely that bariatric surgery and the general anaesthesia that goes with it can sometimes claim lives. But neither Sinead or her parents are listening. As long as the Local Health Authority are prepared to foot the £6,000–8,000 bill, they're more than ready to go for it.

If they are granted their wish, Sinead will need two to six weeks to recover from the procedure, so they're planning to have it done during the school holidays. If Sinead's operation goes to plan, she stands to lose half her excess weight within two years, around half a stone a month, if she opts for a lap band. The gastric bypass alternative would achieve even more, an average of 70 per cent of her excess weight within the same timeframe. But she will need to take nutritional supplements permanently and she will never be able to enjoy normal social eating again.

The main stumbling block to Sinead's gastric surgery, however, is that, according to clinical guidelines, she is not eligible because her BMI is less than 40 and she doesn't yet have any associated medical conditions. Before she can have it done, in order to fulfil the recognized universal criteria she will have to put on *more* weight. Maybe that's why Sinead is standing in front of the hot food counter and helping herself to three very large slices of deep-pan pizza.

The prevalence, accessibility and cheapness of so much high-fat, high-carbohydrate fast food like pizza is half the problem, Poppy knows. The government is attempting to pressurize some of the biggest fast food companies into reducing the calories in their products, but so far to little avail. Why would they want to risk denting their profits by making their offerings less palatable? There is so much more the government could do other than providing bariatric surgery on the NHS to people for whom preventative action is already too late. They could ban vending machines selling sweets and crisps in schools; subsidize fresh fruit and vegetables and fish; levy a fat tax on foods, and re-establish competitive sport as part of the educational curriculum. Annual weight checks on schoolchildren would identify problems early on. In some schools this is already happening. Surely other girls like Sinead could be prevented from getting into the state she is in now? If only she could lose weight the natural way. Why on earth would anyone want to have whole sections of their completely healthy digestive system surgically chopped around just so they can eat less? It's a puzzle Poppy will never understand. And one she doesn't want to think about a moment longer.

Luckily, these thoughts are interrupted by Kirsten, who is behind her and who wants to know whether Poppy has got it on with Ben's friend Frankie.

'Have you even got to first base yet?' she wants to know. It was friendly and well intended but totally out of the blue. Poppy wasn't ready for the question. Base one meant snogging. She suddenly feels herself blushing furiously. What causes that wretched reaction? she wonders. When she asked her aunt about it, Sal said she knew how to deal with it but they'd been so busy talking about other things they'd never got around to addressing it.

Blushing occurs when hundreds of small blood vessels in the face all widen at the same time and suddenly and dramatically increase the blood flow. Poppy is aware of a burning sensation which spreads to her neck and chest, and even more acutely aware that everyone around her must be noticing it too. And that's part of the phenomenon. To be able to blush, that person has to have a sense of how they interact with others. They need a capacity to judge themselves from the standpoint of others, and to be highly embarrassable. You can even make someone blush just by accusing them of doing it. Interestingly, it is something uniquely confined to humans. Even humans who have been blind since birth. No other animal does anything like it. This is because it requires emotional stimuli and sophisticated levels of brain function to trigger it.

Blushing is not a new phenomenon either. It has probably been around ever since the first caveman's loincloth fell off, right in front of his cavewoman's mother. Charles Darwin attributed it to an inherited and exaggerated reflex reaction to the attention of other people. Sigmund Freud rather predictably put it down to repressed sexual excitement and the fear of castration. A more recent explanation is that blushing is a kind of non-verbal apology. The blusher realizes they have done something wrong or naughty in the eyes of others and passively acknowledges this in a way which will generate sympathy and endear them to their social group.

Since Poppy's blushing was brought about by the mere mention of Frankie's name, perhaps Freud was right all along. Maybe in this instance, at least. Repressed sexual excitement is the cause. All she knows, as she is telling herself right now, is that it is a dead giveaway, a visible indication of her true feelings and something she could happily do without. Why is it, she is thinking, that blushing always involves something you want to hide? You want to hide it more than anything else, yet by blushing you immediately confirm it. A strangely contradictory message is being delivered there. A secret is held closely to one's chest but a powerful physical signal draws attention to it. Can Poppy overcome it though? Will she, like her aunt Sal, grow out of it eventually, or is there a medical remedy which can help her right now?

She doesn't know it but one cognitive behavioural technique which can often work is to deliberately try to turn as red as possible whenever she feels a blush coming on. Just as it is almost impossible to make yourself blush when you are alone, distracting the mind by trying to turn red in an embarrassing situation seems to switch off the emotional reflex that causes the blush. For Poppy, in her agony, it has to be worth a go. It's certainly preferable to taking prescribed beta blockers or anti-depressants such as propranolol or venlafaxine, which are occasionally used for the purpose. The main cognitive behavioural technique Poppy currently uses to overcome her blushing is to swiftly change the subject and focus attention on someone else.

'Have you noticed Davina's moustache today?' she asks Kirsten.

'How could I *not* notice it? She looks like the blokes on the 118 118 advert.'

Poppy laughs. 'But she's got sideburns too. And hairy legs.'

'God knows what the bits we can't see look like,' adds Kirsten with a mischievous wide-eyed expression.

'Why doesn't she do something about it? She'd be really pretty if she did,' muses Poppy.

And with that, she and Kirsten grab their salad lunches and go to sit by the window together to discuss the hairy problem of hirsutism in women.

'Seriously, if it was me I'd be bleaching it, plucking it and even waxing it if I had to,' says Kirsten. 'I'd pay for a permanent cure like electrolysis or laser treatment. That would be great for Davina. The darker the hair, the better the response apparently. I wouldn't mind how many times I'd have to go back either. Anything to look less like Rasputin.'

'Waxing would hurt a bit, I guess,' answers Poppy. 'Wouldn't a depilatory cream be kinder?'

'Too right,' says Kirsten. 'There's a new one called Vaniqa. You can get it on prescription.'

'How do you know?'

'Grapevine. My mother, actually. She's going through the menopause and seems to be turning into a gorilla. You know. Wider waist. Unpredictable mood swings. Lots of chest beating and aggression.'

'And lots of thick, matted hair all over her body?'

'So you've met my mum before then?'

Whereupon Kirsten and Poppy fall about laughing over their food.

Hirsutism is a genuine problem for many women. This is especially so in the Western world, where fashion dictates that excess hair growth is unattractive and culturally unacceptable. Usually it is inherited and a common feature in darker-skinned races. But the altered ratio of sex hormones during and after the menopause that favour the virulizing effect of testosterone certainly makes facial hair grow thicker and faster in all female populations wherever they come from. A similar pattern occurs in younger women with polycystic ovarian syndrome, where there are multiple small fluid-filled cysts in the ovaries.

'At least she hasn't got hypertrichosis,' says Kirsten. 'I felt sorry for my mum and looked it up. Apparently, hair grows all over your body. Even in places that don't normally have hair. Like your eyelids and forehead.'

'Around your nipples and kneecaps?'

'And on your shoulders and back.'

'Kirsten?'

'What?'

'I think Davina's got hypertrichosis.'

Chapter 30

Adam, now standing in front of 250 restless sixth-formers, is feeling rather anxious. The school principal is introducing him, much as he expected, as the CEO of a charity specializing in relationship counselling and psycho-sexual medicine.

'Today's event is inaugural,' he drones on, 'a pioneering experiment in adolescent sex education which was originally conceived and developed by the efforts and enthusiasm of the students themselves, and then progressed by the school's pastoral care committee and board of governors.'

Many of the students already look half asleep. Adam, on the other hand, is wide awake. Although his hangover lifted several hours ago thanks to copious amounts of coffee and painkillers, his hands are moist, there are beads of sweat on his forehead and his heart is thumping. He has butterflies in his stomach, his mouth is dry and he is beginning to wish he had taken one more trip to the loo before coming up on stage. He is experiencing all of the classic signs of acute anxiety and, as someone who is usually able to handle pressure, it has taken him rather by surprise. He could easily have taken one of the propranolol tablets his family doctor had prescribed for occasional after-dinner speeches, fear of flying or events such as this. By blocking the effects of his stress hormones, adrenaline and noradrenaline, they completely abolish the physical symptoms of acute anxiety without detracting from mental alertness like other sedatives do. A propranolol tablet would have quickly got rid of all the unpleasant feelings Adam is suffering from now, but he'd been reluctant to take any more medication on top of the concoction he used to cure his hangover.

Suddenly the principal has finished speaking and Adam takes two tentative steps sideways to a position directly behind the rostrum. Two hundred and fifty pupils are watching him expectantly. As far as they are concerned, this guy is a 'sex expert'. This should be fun, they seem to be thinking. Some look smart, bright and interested, appearing attentive and eagerly receptive. Others are less well presented, with ties or collars askew, and are surreptitiously nudging their neighbours with sly little grins on their faces, ready for a laugh. The careful selection process intended to ensure a handpicked mature audience seems to have been less selective than promised. And weren't there supposed to be 200 pupils, not 250? thinks Adam. Things seem to have ballooned somewhat. And today of all days, he reflects, size matters. For his own sake, and that of the sixth-formers, he needs to break the ice. If he is ever going to extract the frank and open questions from the audience that the talk has been designed around, he must introduce a little humour into the proceedings. Humour, he knows, is a fantastically effective way of breaking down barriers and inhibitions. Doctor Ruth, the entertaining, sympathetic and very informative sex-counselling guru of the 1980s, used it brilliantly. Adam uses it himself to great effect whenever he counsels clients. And so he begins.

'At my university, when I was a student, we only ever had one lecture on sex education in five years. Consequently, it was very well attended. But then, our professor of sexology began by expounding his theory that the more sex you have, the happier you are. "Hands up," I remember him saying to us, "all of you students who have sex every day." A number of extremely happy and contented students slowly raised their hands. "There. It fits with my hypothesis," said the professor. "The more sex you have, the happier you are. What about the students who have sex once a week?" More hands went up, but not quite so confidently and the students didn't look as happy as the first group. The third group admitted to only having sex once a month and looked distinctly forlorn. "You see," the professor gleefully explained to us, "frequent sex makes people happier." Then he continued, "Is there anyone here, may I ask, who only has sex once a year?" Immediately, with a banging and crashing of a wooden desk top and a cascading of schoolbooks, one young man leapt up at the back of auditorium, his arms raised in the air and a huge smile on his face, excitedly yelling, "Me! Me! I only have sex once a year!"' At this point in his anecdote, Adam paused for dramatic effect. 'The professor, obviously puzzled to have his prized theory blown apart by this apparent exception to the rule, was almost speechless. "Then why are you so happy?" he gently enquired. "BECAUSE IT'S TONIGHT! IT'S TONIGHT!" screamed the man.'

The students seem to love this and break into spontaneous applause. The chaplain and the observer from the family planning clinic look rather less amused. They are probably wondering where this is going.

You can't please all the people all of the time, thinks Adam, but nevertheless he's already beginning to wonder why he let himself in for this. He knows that, like many people, he usually thrives on the buzz he gets from setting himself exciting new challenges and meeting them. He works best and most effectively with a deadline to beat or when he is juggling many different jobs at once. Like an athlete in competition or an actor on a West End stage, he rises to the big occasion and normally surpasses himself, even in front of a critical audience like this one. It's the same adrenaline rush all these young people might experience on a rollercoaster at the fairground or when watching a particularly gruesome suspense thriller at the movies. Without such challenges, life would be dull, monotonous and unexciting. That is what Adam is trying to tell himself right now anyway. That, and the fact that the rest comes down to personality.

Am I really a Type A personality? he asks himself. Is he as competitive, ambitious and driven as Eve maintains? Does he constantly try to control everything and everybody around him, internalize anger and set unrealistic targets for himself? Is he always rushing around multi-tasking, easily irritated by other people, intolerant of queuing and a poor time-keeper? Perhaps he is. Perhaps that's his problem. Maybe he shouldn't set himself such challenging tasks and worry so much about what other people think of him. It might be better for his long-term health. At this exact moment in time, Adam is only too acutely aware of what others think of him – 252 others to be precise.

'This is a frank and open forum for questions felating to, sorry, relating to relationships, love and sex,' Adam says. Damn! Why do such Freudian slips always have to rear their ugly heads at the most inconvenient times? He can almost feel himself blushing like Poppy does. She'd be excellent at giving a talk like this, he's thinking. Next time, he'll ask her to do it.

'I'm not here to give a lecture,' he goes on. 'Anything we talk about today within these walls is part of a consensual, mutually beneficial two-way thing, as all relationships should be. And what happens on tour stays on tour, so feel free to ask anything at all.' He knows he is mixing his metaphors somewhat, but the pupils seem to get the message about confidentiality at any rate and soon respond. The first question is cautious, safe and completely straight.

'Why does puberty happen at different times in different people?'

Adam explains that puberty usually occurs earlier in girls than boys. He quickly dispels the ripple of triumphalism he notices from the girls, by saying that it sometimes starts as early as age eight.

'There's a growth spurt at around ten and fat is laid down

beneath the skin, especially around the breasts, hips, buttocks and thighs. Hair begins to grow on the pubis, legs and arms. The breasts begin to develop and the dark area around the nipple becomes larger and darker.'

The audience is rapt.

'Puberty in boys kicks in between ten and 17,' Adam continues. 'Usually, the initial growth spurt occurs around 13, when the arms, legs and penis grow, and after looking a bit lanky for a year or two, the rest of the body fills out with more muscle tissue. The later puberty of boys accounts for the fact that they will usually end up taller than the girls. This is because the growing ends of their bones continue to develop for longer, whereas in girls the sex hormones of puberty act on them and stop them growing any further.'

Now it's time for the boys to look rather pleased with themselves. The second question is equally innocuous.

'Can you get AIDS from kissing?'

'It's extremely unlikely that anyone could ever get AIDS from normal kissing.' Adam is careful to insert the word 'normal' as he is aware that some teenagers erroneously see oral sex as a safe alternative to vaginal sex and an erotic form of kissing. He isn't going to embellish his answer, though he will add that even if someone had HIV in their bloodstream, there wouldn't be any significant traceable virus in their saliva, and certainly not enough to transmit the infection. The third question is more probing and probably reflects a personal issue.

'What happens if you are on the pill and you forget to take it for a couple of days?'

'If you take it more than 12 hours after you were due to take it, you have to assume it might not work. The same applies if you are vomiting, have diarrhoea or are taking antibiotics and may not absorb it. Depending on the type of pill you are taking, you should always consult your doctor or contraceptive adviser and use an alternative method of contraception in the meantime.'

'What if it might be too late?' asks the same girl. It sounds to Adam as if she may have had unprotected sexual intercourse herself, presumably rather recently.

'You should still see your doctor or pharmacist. If anyone has unprotected sex, they can still take emergency contraceptive tablets or have a coil fitted, both of which can work even when used several days after the event. The sooner the better though. And if your next period is late, then again you need to talk to your doctor.' Adam anxiously glances over again at the chaplain at this point, and is relieved to see him looking unfazed. It's probably a subject he realizes the students need to know more about. Now the ice has been broken and the questions are coming in thick and fast.

'Why do blokes have erections in the morning?' A titter of embarrassed laughter passes around the room and from somewhere far off one muffled voice is heard to add, 'Only in the morning?' Adam again bats it straight back with a matter-of-fact answer to quell the rising curiosity.

'Testosterone levels in the blood are higher in the morning than at any other time of the day,' explains Adam. 'Together with the stimulation of spinal nerves due to a full bladder, it makes erections more likely to happen in the morning.'

Not to be outdone by the previous direct question on male sexuality, one girl shoots back with one for the females in the room. 'What happens when a woman has an orgasm?'

'Who cares?' shouts the same comedian who interjected earlier. Adam knows that heckling from idiots in the audience is the kiss of death to a discussion like this. He has to nip it in the bud.

'Hang on, hang on a minute,' he says loudly, and with an air of strict authority suddenly in his voice. 'There seems to be someone over there who knows everything there is to know about sex and relationships. Obviously the school's love machine and God's gift to women.' There is a parting of the crowd and a gap opens up around one particular boy who laughs nervously, shifts from foot to foot and looks uncomfortable.

'Would you like to come up here yourself and facilitate the discussion?'

Not surprisingly the boy in question doesn't respond. Adam can sense that the other students are enjoying the chat and want to hear more. He also senses they are irritated by this boy and his rather predictable wisecracks.

'Do you like sex and travel?' asks Adam. The students laugh. The chaplain doesn't know this is a subtle invitation for the boy to f*** off if he doesn't want to join in with the rest of them, but the students do. They seem to appreciate Adam's give-as-good-as-you-get approach, especially as he delivers it with an innocuous smile on his face. The awkward silence is broken by the next enquiry.

'How frequently does the average person think about sex?'

'According to one psychologist, who interviewed more than 4,000 people, men under 25 think about sex once every two minutes and women under 25 once every five minutes. Even with age it remains a constant obsession for many. Our brain, in fact, is the most important sexual organ in the body. It makes us dream about sex, scheme about it, anticipate and remember it. You can be 18 or 80 but your mind is still likely to be inventing all kinds of imaginary sexual encounters and conquests. Sexual fantasies are constantly being created inside your head, pretty much all of the time. But it's normal.'

'You didn't answer my question about female orgasms,' says the same girl as before.

'No I didn't,' responds Adam. And winking in a forgiving manner at the boy he has recently embarrassed, he adds, 'And unlike *him*, I do think it's an important one. In the early stages of sexual excitement, the blood vessels around the female genitalia dilate and fill with blood, making the whole area increase in size and feel fuller and firmer. The clitoris enlarges and emerges from its protective hood. The breasts swell in size, sometimes by as much as 20 per cent, and the nipples can become more erect. To become fully aroused though, it can take up to 20 minutes for these natural processes to happen, including the production of enough lubrication for sexual intercourse. As the sexual tension increases, the muscles throughout the woman's body begin to tense up and contract. Breathing can get quicker and the skin becomes flushed. Some women say that they can feel the muscles in their vagina lift or pull up.' Adam is pausing for dramatic effect at this point, and he can tell that it's not just the pupils who are spellbound but the family planning nurse and the chaplain as well.

'When a woman has an orgasm, her pulse speeds up and her pupils dilate in size. Some women might hold their breath at this moment. From the entire area of the pelvis, vulva and clitoris, sensations of rhythmic contraction will occur, coming in pulses about every second. She might experience anything between three and a dozen spasms, which gradually become less frequent and less intense. The contractions slowly fade away and most women wouldn't be able to identify exactly where they were coming from.'

The audience, both male and female, is entranced. You could hear a pin drop in the hall, such is the concentration. Adam has deliberately given a very full answer, partly because Mr Truculent had made so light of it, and partly because Adam's talking about a subject that fascinates him.

'But isn't it true that about 30 per cent of women never reach orgasm at all?' says the same girl. This girl is going places, thinks Adam.

'Yes, that's true. For many women, orgasm can be a very elusive thing. Emotional security, privacy and slow, patient erotic build-up from a skilled and experienced partner all play their part. But with experimentation with what works best for you, anything is possible.'

Adam is tempted to ask if his answer has totally satisfied her. But on reflection the terminology sounds rather suggestive so he thinks better of it. The inquisition goes on.

'Is masturbation bad for you?'

'No, not at all. For men and women alike, it's a normal activity. It's both a natural form of sexual relief and a triumph of stimulation over inhibition. It's also one of the best ways that a woman who finds it hard to reach orgasm can learn to get there.'

'What about porn?' someone then asks from the periphery of the room. The questions are becoming much less inhibited. And more open-ended.

'Does anyone know where the word pornography actually comes from?' asks Adam. Nobody does.

'Pornography comes from Ancient Greek. It literally means "the writings of prostitutes". For centuries, human beings have used visual images as powerful stimulants of sexual arousal and of sexual fantasy. In that regard, it can be therapeutic for people whose libido needs a boost, or who desire an extra bit of imaginary excitement. On the other hand, it can portray women as sex objects whose only purpose is to satisfy the carnal desires of men. It can exploit, dehumanize and degrade people, both women and men, and encourage inappropriate or violent sexual behaviour which is not consensual. Wherever sexual practices are not welcome or without love, relationships suffer.'

'Do you watch porn?' the same inquisitor asks. This really makes the audience sit up. Even Adam is surprised at this one, and clearly, from his renewed and obvious interest, so is the chaplain. Adam feels a tiny surge of anxiety that makes adrenaline course through his veins.

'Do I watch porn?' he echoes. 'Do I watch porn? Do *you*

watch porn?' He is frantically thinking on his feet. 'Does anyone here watch porn? The question and the answer are irrelevant. It doesn't matter what we do in private. In our own very private sexual world. Provided the activity doesn't harm you or anyone else, it can be seen as normal, sexplorative human behaviour.'

'So you do watch porn,' shoots back the questioner to another ripple of slightly shocked laughter from the audience. Clearly, Adam's previous attempts to curtail such difficult and searching questions have not been as successful as he would have liked. He is suddenly reminded of the physiological fight or flight response that subconsciously urges him to finish the discussion right now and flee the room.

In Neanderthal man, the fight or flight reaction prepared him for the ultimate physical challenge. A series of nervous and hormonal reflexes of lightning speed would enable him to either stand and fight a wild animal or flee any danger as fast as he possibly could. To achieve this, a complex series of biological changes would automatically occur. His heart would beat faster and his blood pressure would rise. His breathing would quicken and his reflexes sharpen. His muscles would fill with blood and his skin would cool and sweat. Extra glucose would pump into his bloodstream and his mental alertness would be heightened, honed and focused. His fighting ability would be enhanced to a degree that would be impossible during any moment of relaxation or unguardedness. Even if he turned to flee his aggressor, the extra oxygen in his lungs, the blood in his muscles and the increased circulation pumping around his body would enable him to run more swiftly and nimbly than at any other time.

The fight or flight reaction is, in short, a lifesaver and it is every bit as vital and effective today as it was at the beginning of time. If an out-of-control juggernaut careers off the road and on to the pavement where pedestrians are walking, they scatter with amazing speed. If an unruly mob of drunken football hooligans rushes towards us waving broken bottles, we escape in the opposite direction exceedingly fast. But if we're cornered like a proverbial wounded beast, we will fight for survival and the fight or flight reaction will protect us, defend us and minimize pain to keep us alert and alive. There have even been substantiated reports of people exhibiting superhuman agility and performing Herculean feats of strength when acting under duress. A mother was able to lift a car off a toddler unaided. A man leapt a 12-foot wall following a bomb explosion. A farm worker walked several miles for help after his arm had been severed by the blades of a combine harvester.

Adam isn't facing the rotating blades of a combine harvester, however. He's facing something much more immediate and threatening: the question about whether or not he watches pornography.

'Actually, I don't,' he says, rather more curtly than he'd intended. 'To me, it's a rather voyeuristic pastime,' and then, realizing that for the first time he'd been a trifle judgemental, he quickly adds, 'but that's not to say that there's anything necessarily wrong with that.'

'So you'd rather be a participant than a spectator,' someone concludes.

'Nicely put. You could say that,' agrees Adam, 'although we're not really here to talk about me, or any other individual for that matter. Unless I have a volunteer?' he adds, looking at the latest would-be torturer. And still the questions keep coming.

How many calories in seminal fluid? Answer: seven. How many sperm in an average ejaculate? Three hundred million. How long is an average man's penis? The average Western male has an erect penis measuring 15.25 centimetres. What do you call a fetish for high heels? Altocalciphilia. What is frottage? When someone rubs against you in order to become aroused. Apparently very popular at rush hour on public transport and generally perpetrated by perverts. There were questions about the difficulties of first-time sex, questions on whether it's true that you can't get pregnant standing up. Questions on virginity, oral sex and wet dreams.

Usually, Adam would handle this barrage of interrogation with ease. But increasingly, despite his best efforts to modify and deflect the most sensitive questions, the tension is getting to him. Why, when his Type A personality normally responds to a challenge, is he losing his confidence and floundering now? It's all a matter of degree. Adam is well aware of the difference between good stress and bad stress. When stress is handled well, a person feels rewarded, satisfied and effective. They feel mentally attuned to making decisions, more creative and aware. A person feels confident and in control of the circumstances in which they find themselves. This is therefore good and fulfilling stress. It's also healthy stress because it's associated with the level of adrenaline which makes us feel empowered, but which does us no long-term damage. Without any stress or challenge whatsoever, we would feel frustrated and bored. In a work situation, we would feel undervalued and unappreciated, and this in its own right can be very stressful. Too *little* stress can be associated with a range of unpleasant and stressful emotions. But there comes a time for all of us when unremitting and high levels of stress become overwhelming and intolerable. We feel overloaded, anxious and irritable. We start making mistakes, feel more self-critical. Eventually, we feel exhausted and burnt out.

Recognizing the growing tension within himself, Adam decides he should try to reintroduce a little humour into the proceedings. Ambiguous sexual orientation questions, questions on 'coming out' and fetishes are weighing him down.

'We've talked a lot about sex itself,' he says, skilfully changing tack. 'But what about the *relationship* side of sex? Sex without feelings and emotions, sex without affection or love for your partner is merely a mechanical act. You might just as well be having solo sex. Does anybody have a question about relationships?' After a short pause, Adam gets the cue he's been waiting for. A boy is suggesting that relationships never last anyway.

'Isn't that why there are so many divorces?' he asks. 'Doesn't the sex just become too mundane and routine, and the couple get fed up with each other?'

'Oscar Wilde would have agreed with you,' Adam says. 'He

famously said: "Polygamy is having one wife too many, and monogamy is the same." Of course, Oscar Wilde, being gay, wasn't too interested in having a wife at all. The thing is, any relationship has to be worked at and nurtured,' answers Adam. 'It takes both partners to compromise and negotiate with one another. Including in sex,' he adds. The chaplain, who was nodding approvingly a moment ago, now suddenly stops. A glance at his watch tells Adam that he is expected to talk for another 15 minutes or so. Why is it that time seems to pass so much more slowly when you don't want it to?

Subjectively, the passage of time really is different to objective time as measured on an accurate clock. Time seems to speed up or slow down depending on what we're experiencing. If we're really busy and our brain cells are firing away rapidly, the more images and information our brain has to process. More is done within the same timeframe and this gives the impression that time is genuinely lasting longer. The firing of brain cells is controlled by chemical messengers called neurotransmitters. Excitatory ones speed them up and inhibitory ones slow them down. Adam's excitatory ones are at their peak right now, and suddenly he comes up with a brainwave.

'As far as relationships go, some men seem to prefer their *dog's* company to that of their wives,' Adam suddenly says. And once again the audience is silenced.

'Think about it. What do these men say? Comparing their dogs to their wives, they say that the later they are when they come home, the more excited their dogs are to see them.'

The students are smiling.

'Dogs don't notice if you call them by another dog's name.'

More smiles.

'Dogs actually like it if you leave lots of things on the floor, and dogs find it genuinely amusing when you've had a lot to drink. If a dog has babies you can put an ad in the paper and give them away.'

Some laughter.

'And if a dog smells another dog on you, they don't get angry, they just think it's interesting.'

The students are clearly enjoying the extended joke.

'And finally,' adds Adam, 'if a dog leaves, it won't take half your stuff,' to a round of spontaneous applause. Adam then asks if there are any further questions on relationships.

'When would Oscar Wilde have known he was gay?'

This is a question neatly phrased to relate back to the previous one. It could also reflect concern from the person asking it about his own sexuality.

'It depends. Some people know instinctively from a very early age whether they are gay or straight. In general though, it seems that feelings of sexual orientation start at around the age of 14 for boys and 16 for girls. That may seem strange because it might be at least another two years before any of them have their first sexual encounter.'

Adam pauses to let this sink in. He knows that about one in 20 of the teenagers in the room will have a homosexual relationship either soon or at some time in the future.

'The thing is, at this age, it's physically and emotionally confusing to work out exactly who you are and who you are

attracted to. It's not abnormal to develop a crush on someone of the same sex. Sometimes it's a passing phase.
Sometimes it leads to a same-sex encounter. Sometimes
that'll feel nice and be the first of many, sometimes it will
just be something you put down to experience. Sexuality
takes a while to develop and settle down. That doesn't
matter. In time, you'll all know what your true feelings are,
who you are attracted to and what relationship you want to
be in. And, by the way, what other people think about that is
utterly irrelevant. Be yourselves. And be proud of it.'

'What is love?' someone quietly asks from the front row.
The question comes from a roly-poly little bat-faced girl
who reminds Adam of a line in the Simon and Garfunkel
song 'You Can Call Me Al'. Only this girl has ponytails as
well. And there's something about her that suggests vulner-
ability, that she hasn't known real love, or has experienced
the wrong kind of it. Then again, maybe she just has a crush
on one of the boys and is living out her own fantasies.

What is love? Adam muses. There's a question, and looking at this innocent girl, one that deserves a good answer.

'What is love?' Adam repeats out loud and takes a deep
voluntary breath before answering.

'Love,' he says, 'is a highly complex thing involving friend-
ship, commitment, romance, intimacy and yes, sometimes
sex as well. Sex doesn't *have* to come into it, because there are
many different kinds of love. Think about it. There's mother
love and puppy love. There's first love and true love. There's
baby worship and hero worship. There are love objects and
love play. There are love letters and love songs. There are love

affairs and lovemaking. There's also tenderness, affection. Yearning. Eroticism. You might think it's easy to fall in love, and that love is for ever. But love also has the capacity to cause pain. So there's also love-hate and love-sickness. Broken hearts and bitter-sweetness. Love, as you can see, is not as simple as you might think.

'Why does it happen though? Romantic love, which is what most people mean when they talk about love, has a survival value. It brings about the bonding of two people, which is a perfect scenario for having babies and then caring for them and protecting them. Without love, sex can be a meaningless act. It is already disconnected from the purpose of reproduction in humans, as we're the only species to indulge in it whenever we choose, rather than when the oestrus cycle determines that the female is ready. We're the only species that can have sex all year round. Yet as I say, sex and love are not the same thing. There's also the love you find in friendships and in social networks. These give you a sense of attachment and belonging. These are important for well-being, self-esteem and good health.'

'Can you be addicted to love?' someone shouts.

'In the sense that some people seem to fall in and out of love very easily and very quickly, no. Some people have personalities which crave the warmth and attention of other people. They're social butterflies and need company. But these temporary attachments to people are superficial. They're not addictive. It isn't the same as addiction. But interestingly, there is a neurochemical basis for the power and obsession of deep romantic love. Both phenylethylamine and dopamine are neurotransmitters in the brain which act on the limbic system and cortex to create intense emotional pleasure. This is the same neurological pathway which allows addictive substances like nicotine, heroin or cocaine to activate the dopamine reward system. So in that sense, love, and the intense emotions that come with it, can sometimes be addictive. And that's why the ending of a love affair can be so heartbreaking. It's like a drug addict going cold turkey. And sometimes equally dangerous.'

As Adam says this, he looks at the girl who asked the question in the first place and he notices the ghost of an unreadable expression pass behind her sad eyes.

'That's why,' he quickly adds, 'we all need to look after each other and support and understand each other whenever that happens.'

This was the basis of his job as a counsellor, Adam thought to himself – the ability to read people's expressions, to interpret the real meaning of their questions or answers and to help them towards solutions to their problems, which he is at pains to point out are never unique.

'Is there such a thing as *sex* addiction?' pipes up the next questioner.

'You read about sex addicts in the newspapers, don't you? Russell Brand. David Duchovny. Michael Douglas. Tiger Woods. But do they really suffer from hypersexuality, or is it just a convenient excuse for men behaving badly? Or women behaving badly, for that matter? Sexual urges and behaviour and thoughts which are extreme in frequency or out of control are associated with people who have addictive or obsessional personalities. When it's genuine, there is usually a problem with low self-esteem, and self-destructive behaviour often fuelled by alcohol or other recreational drug use. Sexual compulsivity and addiction, however, contrary to what you might think, isn't much fun. The difference between someone with a high sex drive and sexual addiction is that an addict suffers from all the negative feelings of guilt and shame that goes with their addiction. They live double lives. Their relationships never last. They get into legal problems. They pick up sexually transmitted infections, and they never achieve happiness through their behaviour. The life of a genuine sex addict is a hollow, pretty unrewarding one.' There is a pause as students take this all in.

'Can you pick up a sexually transmitted infection from a toilet seat?' another voice asks at the back.

'The short answer is no. It doesn't happen. The longer answer is yes, it was once recorded in medical literature. About 100 years ago. A young girl with a recent cut on the back of her thigh sat on a public toilet seat where a man with an open syphilitic ulcer had recently sat. The bacterial spirochaetes which cause syphilis had survived long enough on the toilet seat to find their way through the girl's wound into her bloodstream, and she contracted the disease. Proof, if you like, that a sexually transmitted disease is not always sexually transmitted.'

'What's the best way to avoid a sexual infection?'

'Condoms. Or abstinence. Being in contact with anyone else's bodily fluids can potentially pass on a sexually transmitted infection. There's been a huge rise in the numbers of STIs affecting the 16–24-year-old age group, so if and when any of you decide to become sexually active, you need to take precautions. Not just against sexual infections but against unplanned pregnancy. Condoms provide protection against both.'

'What if a condom breaks?'

'Condoms are tested rigorously. They can be blown up and stretched on machines to an enormous degree before they break. When they fail, as they do in about 2 per cent of cases, it's usually because they're allowed to slip off at the wrong moment rather than them breaking.'

The students are still loving the discussion and clearly learning quite a lot. Adam is pleased, as he'd much rather they learned reliable facts than picked up nonsense in the playground or common room. There are only five minutes left of his talk, and his stress challenge is almost over. Stress, he thinks to himself. Hasn't he got enough of it in his life already without inflicting yet more on himself? A baby on the way. A large mortgage. Hassles at work. Ageing, ailing in-laws. Rebellious teenagers. Too much drinking and too many road rage incidents. It must all be taking a toll on his health. How long was the list the doctor had given him of medical conditions caused by stress? Is there even a single organ in the body not adversely affected by excessive stress? It is time, he tells himself, to wind up proceedings.

'Right,' he says in a loud and commanding voice. 'A couple of minutes left. We can't finish on such serious notes as sexually transmitted infections and slip-ups with condoms. Or slip-offs for that matter. Does anyone have any final thoughts? Anything at all?'

'If you're a sex counsellor, Mr Enniman, does that mean you're really, really good at sex yourself?'

Laughter courses around the room. The question has all the hallmarks of a pre-planned strategy.

'I wouldn't be the person to ask,' says Adam, skilfully avoiding the obvious trap. 'You'd have to ask my wife.'

'I already did,' says the cheeky young man, who has the audience's attention. 'She says you're not as good as me.'

Now the auditorium is in uproar and Adam too finds himself laughing. He has to admit he walked straight into that one.

'I have a real problem with sexual harassment,' says a girl quite seriously.

'And what problem is that?' asks Adam.

'There's not enough of it.'

More laughter. One particular group of students seems to have prepared for a grand finale.

'What do you think about the side effects of the new Viagra cream they've just brought out?'

Adam didn't know there was a new Viagra cream.

'And what would they be?'

'You just rub the cream on, but you can't bend your fingers for the next three days.'

This one brings the house down. Before Adam can make himself heard again, up pipes yet another voice.

'We were told you might be returning to give a lecture on premature ejaculation. Will you be coming soon?'

By now, Adam has lost all control of his audience. It is a

neat, witty joke to end on, and Adam doesn't mind playing the stooge. Suddenly at that moment, as if by magic, a rather flushed and stern-looking principal appears at his side and rapidly winds up the catechism.

'Well done,' he says without much real enthusiasm, as he shakes Adam firmly by the hand. 'Very informative. Very . . . entertaining. Perhaps you have, er . . . a colleague who might grace us with their presence next year?'

Chapter 31

UNLIKE HER HUSBAND, who is constantly chasing his tail, setting himself unrealistic deadlines and time-watching, Eve is cool under pressure, organized and adept at delegating work to competent others. She has arrived in good time at the foundation hospital where she is due to interview the consultant psychiatrist, and the film crew have already set up under the watchful and rather suspicious gaze of the hospital's head of PR. The last time a film crew came here, they completely stitched up the primary care trust by editing their interview in a way which clearly suggested a problem with their infection control. Considering that the broadcast was watched by over 2 million people, the hospital were understandably very aggrieved. The PR lady was not going to let that happen again. It wasn't Eve's TV company who had behaved irresponsibly but it had taken a great deal of patient persuasion on her part to get them to agree to an interview at all. The subject was the influence of food on mood. No, she had insisted, the interview would *not* be touching on the topic of rude food. Eve had had to reassure Eleanor, the PR girl, on that point several times. Not *rude* food, only *mood* food. Everything above board.

Before Eve was allowed to begin the interview, she'd had to thoroughly wash her hands with the alcohol gel provided from a dispenser on the wall. It was a routine she was only too pleased to follow, just like the rest of the crew, especially since her own mother had recently had to spend several weeks longer in hospital after a simple procedure because of a nosocomial infection picked up in her own local hospital. Eve knew increasingly nasty antibiotic-resistant bacteria were everywhere these days.

There are between 2 and 10 million bacteria between our fingertips and our elbows. There are more germs under our watches, bracelets and rings than there are people living in Europe. Some 5.5 million people suffer from food poisoning every year. That isn't so surprising perhaps, given that the number of bacteria on your fingertips doubles after going to the toilet, and that 50 per cent of men and 25 per cent of women never wash their hands after paying a visit. That includes chefs and other food handlers. Campylobacter, clostridium difficile, E. coli, salmonella and shigella among other germs can all be responsible for food poisoning. Then there's legionella, staphylococcus and streptococcus – the flesh-eating bug, and all those superbugs in hospitals and out in the community that are increasingly proving difficult to treat. Eve shudders at the mere thought, then gives her hands one more blast with the alcohol gel before she sits down to start the interview.

'How is our mood affected by our diet?' begins Eve.

'Fluctuations in our blood glucose levels can make us irritable and grumpy. If we constantly eat starchy or sugary food, which have a high glycaemic index or, in other words, produce a rapid rise in blood sugar, we quickly secrete more insulin to bring the level down again. The end result is hyperactivity and excitability some of the time, and cravings for sweet things and fatigue at others.'

'But it isn't just about the energy density of food, is it?'

'Not at all. Many of the influences on our moods and emotions are neurotransmitter-based. The food we eat can be broken down in our bodies and converted into these neurochemicals, increasing the amount of serotonin, dopamine and acetylcholine within our brain, and producing significant mood swings.'

'What about artificial colourings and preservatives? What role do they play?'

'Increasing evidence suggests they are absorbed very quickly through the stomach wall, crossing the blood–brain barrier and contributing, in some people at least, to attention deficit hyperactivity disorder or ADHD. In those with food intolerances, an abnormal permeability of the gut can allow a number of unfamiliar foreign proteins from the food chain to gain access to our bloodstream and produce a whole range of psychological symptoms as well as physical symptoms.'

'Are there any foods which are particularly likely to cause problems like this?'

At this point the psychiatrist looks up and to the right as

if deep in thought. It's a common reflex reaction which people do all the time. It makes Eve wonder why it happens and whether there's any mileage in doing a programme on it.

In truth, there isn't. Scientists don't really know. The theory is that we use different senses to help us recall certain memories, images and thoughts. We might remember things visually or through sound, for example. So when the consultant recalls the information he needs before he answers Eve's question, he looks up and to the right to recall the visual image of the erudite paper he'd published on the subject. Different specific eye movements, it is postulated, call into play different senses. Touch. Sound. Taste. Eve's thoughts are interrupted by the professor's answer.

'That depends on the person. Someone with a tendency to migraine, for example, can have an attack triggered by consuming Chianti, cheese or chocolate. These contain tyramine. Caffeine and alcohol can also promote attacks.'

'So are there specific changes we can make to our diet that would help someone with, say, anxiety or depression?'

'Absolutely. But it isn't just a case of avoiding culprit foods. Sometimes we need to consume more of the foods that might help us. I'm talking about food as therapy. Deficiencies of certain foodstuffs can lead to all kinds of health problems. If you take your poorly dog to the vet, the first question he asks is, "How are you feeding him?" Vets understand the relationship between animal food and animal health. But us humans are no different. Deficiencies in omega-3 fatty acids can lead to depression. More turkey, fish, chicken, wholegrains, nuts, seeds and peanut butter can combat it. Lack of vitamin B complexes can lead to schizophrenia. Just taking on board more water would make a huge difference to most people because of the way dehydration makes us feel. So more water, more oily fish and more nuts and seeds would significantly help.'

'So tell us more about your Food for Therapy research and the Body Bistro you've set up in the hospital.'

The interview concludes with the consultant explaining the principles of the café he has established for mental health research in the hospital lobby. A special menu for those with premenstrual syndrome or seasonal affective disorder. A different one for those with chronic fatigue syndrome or dementia. A tailored approach for others with anxiety or depression. It's a sort of depression delicatessen and as far as TV is concerned, it's interesting and it's a first. It's also a little bit wacky. And TV audiences love that.

Within 90 minutes of arriving for the interview, Eve has what she needs in the can. As she drives homeward to do the weekly shopping before meeting Adam at her parents' house, she reflects on her last pregnancy, and the last programme she had made on the subject of psychiatry. It was on rare psychiatric disorders. And if she is honest, she would agree it was a bit of a gawp. Rather inevitably, it had started with Gilles de la Tourette's syndrome because of the widespread but erroneous belief that loud foul-mouthed expletives are a typical feature of the condition. In fact, they aren't. Only a minority of people with Tourette's manifest coprolalia, as it's called, and most of the 2–3 per cent of the population who have the syndrome merely exhibit short vocal utterances or physical tics.

The programme also talked about Capgras syndrome, which is the delusional belief that a person close to you, usually a wife or husband, has been replaced by an exact double or impostor. It featured De Clérambault's syndrome, or erotomania, where the sufferer holds the unshakeable belief that another person is secretly in love with them. It showed clips of *Fatal Attraction* and *The Bodyguard* to illustrate it. It described Munchausen's syndrome, the persistent simulation of illness in order to achieve frequent hospitalizations, and Munchausen's syndrome by proxy, a variant involving a caregiver repeatedly seeking medical attention using non-existent symptoms or deliberately induced injuries in a person who is dependent on them as an excuse. It featured the case of the nurse Beverley Allitt as an example. There was Ekbom's syndrome where there are delusions of skin infestation with an insect, parasite or, more recently, nanoprobes, and Folie à Deux, where the mad delusions of one partner in a relationship are gradually transferred to the other, less dominant partner. There was also Amok, a dissociative disorder described in South East Asian men, who after a period of social withdrawal exhibit frenzied and violent outbursts resulting in injury, suicide or murder. Hence the title of Eve's documentary, *Running Amok*. It had been an all-time runaway success. Madness.

Eve wants her next project to be about palpitations. She often experiences them herself and wants to learn more anyway. Producing a programme would guarantee her special insight into the topic and introduce her to all the right people. It is a perk of the job she can now enjoy as a senior producer and a far cry from the tedious political interviews she used to have to cover which never saw the light of day anyway. Palpitations is a good meaty subject. She'd just have to avoid calling it *Heartbeat*, she told herself. What she'd discovered already was that in the UK, irregular heartbeats affect at least one in four people. And, she was surprised to learn, they are the leading cause of death overall. Imagine that. They can occur at any age and may lead to sudden cardiac arrest or stroke. Scary.

In one-half of the victims with the most common type of heart rhythm disorder the condition isn't even detected in advance of a stroke and around 35 per cent of people who are told they have epilepsy have actually been misdiagnosed. In fact, they have a heart rhythm disorder which brings about momentary but dramatic falls in blood pressure, lack of oxygen to the brain, and seizures as a result. There is even a name for it, Eve had discovered: convulsive syncope (with syncope pronounced with a 'pee' at the end).

There was plenty of material for Eve to fall back on. One hundred thousand sudden cardiac deaths every year. The dramatic example of Fabrice Muamba, the top-flight Bolton footballer who collapsed on the pitch and 'died' for 90 minutes, only to make a full recovery later. TV gold, Eve thought to herself. The programme was just crying out to be made. It would appeal to everyone. Millions with occasional missed or irregular heartbeats. Thousands with palpitations or fluttery sensations in their chest. Others with unexplained loss of consciousness, falls, shortness of breath or dizziness.

Then there were the trigger factors. Stimulants like alcohol, caffeine, nicotine and asthma medications. Nervous system influences like stress, fatigue, prolonged standing and anxiety. Family history was apparently important too. Eve could interview whole families and bring in different age groups. She could exploit people's natural hypochondria about matters of the heart but reassure them by providing a helpline, guiding them to the excellent Arrhythmia Alliance charity. But she'd scare the hell out of them first. Wasn't that how responsible television was meant to work these days? she thought. It was about education, for sure. But most of all these days, it was about ratings. She'd have to make palpitations *sexy*.

Chapter 32

BEN'S PHYSICAL EDUCATION teacher and self-appointed running coach is enthusiastic enough, but totally uninspiring. A mediocre athlete himself in his youth, he constantly tries to live out his unrealized ambitions through his students. And Ben is his prize pupil. Ben, he knows, could well pip the visiting school's favourite for the Wilmslow Mile, Matt Roberts, and retain the trophy for a record second time.

He has always effusively extolled the benefits of exercise. Everyone should enjoy its therapeutic qualities, he preaches, as a healthy body makes for a healthy mind. He frequently lists them. Lower blood pressure and cholesterol levels. Less risk of heart attack and stroke. Fewer blood clots and a lower incidence of cancer. Reduced risk of osteoporosis and diabetes. Better sleep quality and healthier weight. Less anxiety, depression and stress. Boosted self-esteem and improved morale and mood.

He's right, thinks Ben, but in the mood he is in now, he

could happily strangle the guy. When you're nervous and trying to focus on a competitive race that you know is going to push your body to the limit and hurt like hell, the last thing you want to be distracted by are the platitudes of some red-faced windbag with halitosis. Ben isn't running in this race to benefit from the therapeutic effects of exercise, he's running in it because he cannot stand that insufferably arrogant prig Matt Roberts. He'd rather die than let himself be beaten by *him*. If he stands any closer to his PE teacher he probably will die, thinks Ben. The guy has to have the foulest breath on the planet. Either way, Ben needs this race to start so he can begin running very fast in the opposite direction. His dad won't be there to watch, as he usually is; he said he had some sort of lecture to give which could be a bit of a challenge. Frankie, another good runner, is there, although only as a spectator today because he has been sidelined by a dead leg in a rugby game a few days before. A high-impact tackle had crushed part of his thigh muscle against the femur bone beneath, causing the muscle to tear and bleed within its surrounding sheath. It is still very painful and tingly as well as visibly bruised and swollen. Frankie, despite his best efforts in applying the principles of rest, ice, compression and elevation (RICE), can only walk around gingerly with a limp.

Toby is there too, on crutches, to cheer Ben on and has been telling anyone who would listen about his out-of-body experience in hospital. Funny, isn't it, Ben thinks, how he can't just be satisfied with a ruptured spleen, lacerated liver and fractured ankle? Now he has to have flat-lined and nearly died on the operating table as well.

What is an out-of-body experience anyway? Ben wonders. And is it anything to do with a person's soul temporarily leaving their body? Is it always part of a near-death experience, as Toby says it was in his case?

In fact, out-of-body experiences or OBEs can happen without apparent physical or mental trauma. About one in ten people claim to have experienced at least one during their lifetime, although certain physiological conditions make people much more susceptible. Dehydration, sensory deprivation or overload, and the side effects of medicines and recreational drugs can all play a part. The alcohol in Toby's blood together with the anaesthetic during his surgery could well have been involved.

Awareness under anaesthesia, while potentially terrifying, is not actually all that uncommon. Imagine being awake and conscious when the surgeon slices into you with his scalpel. Imagine hearing him share jokes with the scrub nurse about how many holes he's found in your duodenum and how far into the air your blood is squirting from all the little arterioles that have been severed. It must be just like that final scene from the film *Braveheart*, where William Wallace watches his own entrails being pulled out. Yet at least two in 1,000 people may experience it. Thankfully, only about a third of them will feel any pain, and only a minority will be unable to move and inform the medical staff of their plight due to the paralysing effect of the muscle relaxant used.

But no anaesthetist ever wants to use too high a dose of the anaesthetic. In a traumatized patient who has already lost a lot of blood and who may be developing serious breathing or clotting problems, that could be fatal. So it's a balance. Just enough anaesthetic to put the person to sleep and to neutralize sensation, but not enough to affect blood pressure and breathing. The idea is to kill the pain, not the patient. In this situation, an OBE is not unlikely. But, in scientific terms, it is more to do with the workings of the unconscious brain than with any fanciful notion of soul-travel, or spirit-walking.

The parts of the brain involved in imagination and dreams can remain active even when the patient is asleep or anaesthetized. Scientists have even simulated OBEs in patients they have rendered semi-conscious with drugs. This strange phenomenon may boil down to a neurological mismatch between visual and tactile signals. But whatever it is, Toby is obviously going to dine out on his story for years.

How much longer before this race gets started? thinks Ben. Time always seems to pass so slowly when you are tense and anxious. His whole body is geared up for action. His heart is pounding, he is sweating and his muscles are filling with blood as his arteries are dilated by the stress hormone adrenaline. Sugar and fatty acids have been mobilized into his bloodstream for energy as well. It's no different to the physical effects of anxiety that his father, Adam, had experienced when talking to the students about sex and relationships this very afternoon, but in Ben's case the physiological changes in his body will actually prove useful. He will need extra blood in his muscles to supply more oxygen. Including the muscle of his heart. He will need a higher blood pressure to pump the blood around his body more efficiently. And he'll use those extra nutrients in his blood for fuel, which would otherwise, as in Adam's case, be laid down as unhealthy fatty deposits on the inside of his arteries.

This pathological process, the hardening of arteries brought about by the deposition of unhealthy fats and cholesterol called low-density lipoproteins (LDLs), is the basis of heart attacks and strokes. The stress of today's crazy world constantly mobilizes these fats into the bloodstream in preparation for fight or flight. Yet the very reaction which originally evolved as a survival mechanism to protect us has become a mechanism which can kill us. Without regular exercise to burn off these accumulated fats, we develop coronary heart disease and die prematurely as a result.

Many of the runners are stretching their muscles to prevent injury and cramp during the race. Muscle cramps can affect everyone from time to time, and not just in the calves and hamstrings to which athletes are particularly prone. Women can experience uterine cramps at period times, and anyone with irritable bowel syndrome can suffer from intestinal cramps whenever symptoms flare up. Cramp in the small abdominal muscles leads to a 'stitch'. If any of Ben's competitors develop cramp during the race their muscles will not only suddenly become hard and tender but they will have no prospect whatsoever of continuing. But what causes it?

Muscle cramps can occur at rest, say when lying in bed at the end of the day or sitting in a comfortable armchair. But the same conditions which produce cramp then are present during vigorous exercise and to a much greater degree.

As repeated muscle contraction continues, oxygen supply to the individual muscle fibres is reduced at the same time as chemical waste products begin to accumulate. This change in the internal chemical environment of each muscle eventually leads to a powerful and sustained contraction which is vice-like enough to hold a limb in an agonizing fixed spasm that often requires mechanical stretching from some third party to overcome it. Furthermore, for several minutes after cramp, the affected muscle will retain a lower threshold and increased susceptibility to another spasm, which explains why a sufficient period of stretching and massaging to restore the blood supply is very important.

Certain pre-existing conditions can make the onset of cramps more likely. A low concentration of calcium ions in the fluid in which the muscles are bathed. Low sodium levels and dehydration following profuse sweating. Anxiety and the fast, rapid breathing of hyperventilation that accompanies it and which alter the acidity of the muscle environment. All of these make muscles more excitable and twitchy. So it is strange to learn that cramp actually has a protective function. It protects against permanent muscle damage that would inevitably ensue if muscle contraction continued in the absence of oxygen.

The stretching is completed and the boys are now lining up at the start line. Some are alternately straightening and bending their legs to loosen muscles and others are jumping up and down on both feet to keep mobile. The smell of liniment is pervasive but stimulating too, as it's a memorable smell with strong associations to previous athletic endeavour. One of the boys breaks wind exuberantly. Better out than in, he's probably thinking. But not as far as Ben's concerned as he's currently hyperventilating to pre-load his body with as much oxygen as possible while getting rid of as much carbon dioxide as he can. The fart in the atmosphere is untimely and distinctly unpleasant.

Is it an innocuous fart though, as Ben is hoping, containing just hydrogen, nitrogen, carbon dioxide and oxygen, in which case it will be odourless? Or does it contain hydrogen sulphide and methane, in which case it will reek like a bucket of rotten eggs? The way the prevailing wind is blowing, he will soon find out. It isn't exactly the poor bloke's fault, whichever it is, and as usual no one is owning up to it because everyone farts every day, and on average anywhere between 14 and 25 times. Some people, like Uncle Mike, expel up to 2 litres a day, and all of us at any given time have around 200ml – a whole cupful – of gas in our guts. Some people just overeat. Others eat like pigs and swallow too much air. A diet rich in unabsorbable carbohydrates such as beans, peas, broccoli, parsnips, raisins, prunes and apples will make it worse. These foods contain the sugar raffinose, which the human body cannot break down and use. The friendly bacteria in our intestine work on them and produce lots of extra gas. Fructose in sweetened fruit drinks and sodas and sorbitol used as a sweetener in tinned peaches, pears and diabetic foods will also make people fart more. And anyone on a high-fibre F-plan diet will soon learn what the letter F actually stands for.

The same culprit who broke wind a moment ago has just

done it again. A modern-day 'Le Pétomane' if ever there was one. Maybe he should be running in Shreddies underwear, Ben is thinking. This features a specially activated carbon back panel that absorbs the flatulence odours. Ben had been amused to read the advertising bumph on it recently and remembered it saying that due to its highly porous nature the odour vapour became trapped and neutralized by the cloth, which is then reactivated by washing the garment in water. Apparently, this activated carbon cloth is the very same stuff used in chemical warfare suits. How appropriate, thinks Ben, that's exactly what he is dealing with.

And then suddenly the starting gun fires and the runners are off. Jostling for position near the front of the field on the inside of the first bend, Ben is aiming for a 60–65-second first 400m lap. That's pretty fast, but not as fast as Usain Bolt. The Olympic sprint gold medallist and the fastest man on earth runs 100 metres in just under 10 seconds. Ben, however, worships a different idol. The winner of the Olympic marathon runs every 100 metres of his 26.2-mile course in about 16 seconds. He maintains that fantastic speed for over two hours. He keeps going for another 105 laps after Usain Bolt has finished. That, in Ben's mind, is even more staggering, and the pinnacle of athletic achievement. If he can even begin to emulate that and complete this mile in around 4 minutes 40 seconds, he'll be delighted with himself. Especially if he beats his main rival, Matt Roberts. As it is, Matt is running just ahead of him and looks strong. Boris, another potential threat, is struggling at the back of the pack with a stitch.

This sharp upper abdominal pain means he has either

eaten too close to the start of the race, or the muscles of his diaphragm are being affected by the build-up of lactic acid. The changes in the human body during vigorous exercise are considerable. Normally at rest we take between 12 and 20 breaths every minute, and inhale and exhale about 11,000 litres of air every day. During those 24 hours, we lose about half a litre of water vapour in our breath, and this is the mist which forms on a panel of glass when we breathe on it. We extract as much oxygen from the air that we breathe as we can, and the tiny air sacs within the parenchyma of our lungs, the alveoli, transfer it into our bloodstream. At the same time, carbon dioxide, a waste product of metabolism which circulates in the blood, is excreted into the alveoli and from there into our breath. This process is known as gaseous transfer, and in top athletes has been found to be abnormally efficient, giving them the extraordinary talent they have, and an ability that no amount of training could achieve in any ordinary mortal. At the speed Ben and his competitors are going, the rate of breathing can rise to 60 or more breaths per minute. This reduces the rising amount of carbon dioxide dissolved in the blood from muscular activity. The excess carbon dioxide lowers the pH of the blood and makes it more acidic. As the acidity increases and the cerebro-spinal fluid is lowered in turn, the brain senses this. The rate of respiration is increased to allow more carbon dioxide to be removed via the lungs.

Contrary to popular belief, it is not the level of oxygen in the blood that determines the rate of breathing but the level of carbon dioxide. You'd have to be near death before the very low oxygen level would make a difference. Even as it is, there is a delay between the carbon dioxide level increasing and the breathing rate becoming faster. That is why it takes half a minute or so between running to catch a bus and actually getting short of breath once you're sitting on your seat inside it.

Our two lungs are slightly different sizes. The left lung is smaller than its twin, to accommodate the heart. It also has two lobes instead of the three on the right side. If you could join all of the respiratory tubes together end to end, they would stretch for 50 kilometres, and the 300 million alveoli have a combined surface area equivalent to that of the centre court at Wimbledon. Together, they contain some 4–6 litres of air and even after fully breathing out there is still a whole litre of air left behind. It is known as the 'dead space'. This litre of air remains parked in the larger airways, where no gaseous transfer can occur, and if you were to artificially extend it by using a long snorkel or pipe, for example, you would have to breathe even more deeply to enable any new oxygen-rich air to reach the alveoli. If you couldn't do this, deoxygenated air would simply move up and down in the windpipe and you would eventually suffocate. The alveoli are efficient for oxygen transfer but so tiny that surface tension would normally make them collapse. A special lubricating secretion called a surfactant reduces this surface tension and keeps them open. Once the oxygen has reached the bloodstream through the alveolar walls, it has to be transported back to the heart through a network of capillaries. These capillaries if connected end to end would measure 1,600 kilometres in length, yet the passage of blood from the heart through the lungs and back again takes just six seconds.

As Ben embarks on the third lap of his four-lap race, his heart rate has increased to 190 beats per minute. At rest, it muddles along at around 48 beats per minute, and when he is asleep, plummets as low as 40. The reduction is because there isn't much demand during sleep for oxygen in the inactive muscles of his body. The average resting heart rate is around 70 beats per minute but it is faster in people who are unfit because their heart muscle is less toned and efficient and has to compensate for a lower cardiac output with each beat by speeding up. Conversely, people like Ben with a strong powerful heart can have heart rates as low as 35–40 and still function normally. The heart muscle, just like any other muscle in the body, will enlarge and become more powerful with regular exercise. Chest X-rays can demonstrate this visually. The heart of a trained athlete will fill half the diameter of the chest, much more than the heart of someone sedentary. But the resting heart rate can change throughout life.

In a foetus, the heart starts pumping at as early as four weeks and beats at about 150 beats per minute. At 12 weeks, the stage Eve's unborn baby is at now, it is pumping out as much as 60 pints of blood per day. At birth, a newborn baby has about one cupful of blood in its circulation compared to around 8–10 pints for a fully grown adult. Ruby, at five, will have a heart rate about twice as fast as an adult's and a heart roughly the same size as her fist. The resting heart rate can

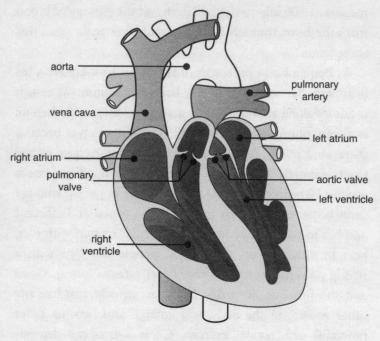

aorta

pulmonary artery

vena cava

left atrium

right atrium

pulmonary valve

aortic valve

left ventricle

right ventricle

The Heart

also fluctuate dramatically due to illness. When abnormalities of its own unique electrical wiring system develop, the heart rate can vary between 30 and 300bpm. In 'heart block', the regular signals from the body's natural pacemaker (the sinoatrial node) in the right atrium are not conducted downwards to the muscular ventricles properly, and the heart then beats abnormally slowly. In supraventricular tachycardia, chaotic and multifocal signals are sent out from the atria and the heart can beat abnormally quickly.

The heart is a hardy, reliable and extraordinary organ. At

its average healthy rate, it beats 70 times a minute, 100,000 times a day, 35 million times a year and 2.5 billion times in a lifetime. Weighing less than a pound, at around 11 ounces, it pumps 2,000 gallons of blood through 60,000 miles of blood vessels every day and does it with the force of a tennis ball being tightly squeezed in your fist. Of all the muscles in the human body, despite being small the heart does the greatest amount of physical work. With a power output of anywhere between 1 to 5 watts, its 80-year output at an average of 1 watt amounts to 2.5 gigajoules. That's a daily energy output sufficient to drive a truck 20 miles, or to the moon and back over a lifetime. It pumps blood into the aorta at 1 mile per hour but along all of the capillaries of the body at a mere 43 inches per hour to allow sufficient time for the transfer of oxygen and carbon dioxide to and from the tissues of the body. In 8 seconds, blood circulates from the heart to the brain and back. In 16 seconds it travels to the toes and back. In time it reaches each of the body's 75 trillion cells except those of the cornea, which uniquely obtain all the oxygen they need from the air with which the surface of the eye is in contact.

According to conventional wisdom, the circulation was first demonstrated by William Harvey in 1616. He was reputedly the first to state that there was a finite amount of blood in any one person's body, and that it flowed in only one direction. This must have been a bit of an insult to the tens of thousands of shepherds and farmers over previous centuries, who had butchered their animals for meat and observed on a daily basis the blood squirt from their bodies until it stopped of its own accord. But nevertheless, William Harvey took the credit.

The circulation does indeed flow in one direction. From the heart, in pulsatile arteries to the rest of the body and lungs, and from there, much more slowly back to the heart in veins. Five per cent of the blood coming out of the heart supplies the heart muscle itself via the coronary arteries. Twenty per cent goes to the brain and central nervous system. Twenty-two per cent flows to the kidneys. The rest is circulated to all the other tissues of the body, including the muscles.

This is exactly where Ben needs it now. When he hears the bell that tells him he is entering his final lap, his lungs are burning and his muscles are on fire. The demand for oxygen is enormous and the dull ache mounting in his calves and thighs is telling his body he is in the physiological state known as oxygen debt. His muscles are actually using more oxygen than his heart and lungs can supply. He is therefore creating energy anaerobically through the chemical combustion of carbohydrates in his body, and producing a build-up of lactic acid in his muscles in the process. This accumulation is causing the increasing pain in his calves and thighs and would, in anyone less fit, make his muscles intensely stiff the next day.

Matt Roberts is still just ahead of him and has increased his pace. However, he is not pulling away. Ben is just on his outside shoulder and feeling comfortable. With 300 metres left to run, getting sufficient oxygen on board will not be a problem. It is a sprint now, and the oxygen debt can be repaid later. Ben knows the body will always collapse first before the heart gives out. No one is going to die. It just feels like it. Both athletes are equally desperate to win and there is no question of superior psychological motivation on either side. Now with 200 metres and one last bend in front of them, it is all coming down to their muscles. Who will prevail and who will trail in second?

All the muscles they have now, they were born with. No amount of regular exercise on the track or in the gym can increase their numbers, they can only increase the size of each individual muscle cell. Forty per cent of their body weight is composed of muscles and there are at least 640 individually named examples of them. The biggest is the gluteus maximus in the buttock and the strongest, pound for pound, the masseter or chewing muscle of the jaw. This is not quite as strong as the hyena's masseter, which is capable of a bite force of 2 tons per square inch, but very powerful all the same.

The two boys' leg muscles are propelling them forward now in a headlong rush. But their other muscles and the muscles in every human being's body can do so many other things. Make them speak. Help them swallow or digest their food. Enable them to blink, breathe or empty their bladder. Seventy-two muscles are responsible for speech and articulation, and 30 for facial expressions such as surprise, sadness or joy. Seventeen muscles are required for smiling and 42 for frowning. Yet muscles can only work by contracting. Attached to the skeleton at either end (except for the tongue which only has one attachment), muscles can move bones and create movement. Contrary to popular belief, muscles can't push, they can only pull. The way the body gets around this is to have agonist and antagonist muscles which do equal and opposite things. The triceps oppose the biceps, the quadriceps oppose the hamstrings, and so on. Contraction is achieved when tiny microscopic fibres in individual muscles slide past each other like the opening and shutting of sliding glass doors. This process uses chemical energy from food together with oxygen and 1lb of muscle will burn 75 to 100 calories of energy each day even at rest. Ben's leg muscles are burning a lot more energy than that right now. And his brain and nervous system are making them do it voluntarily. He could stop running any time he wanted to, and it is tempting. But not until he passes his hated rival, and breaks the finishing tape.

As he emerges from the final bend, he makes his move. Matt isn't really sure where Ben is, or how far behind. He can only concentrate on his own running and his own pain. He feels confident about making his surge for the line and is ready to crank it up. At that precise moment, however, Ben appears at his shoulder and overtakes him. No matter how hard Matt tries, he can't make up enough ground. Ben wins in a personal best time and retains the Wilmslow Cup for a record second year.

Matt lies flat out on the cinder track, exhausted, while Ben is euphoric, jumping up and down celebrating with his mates. This is the psychological difference between winning and losing. The pain-numbing joy of triumph versus the agonizing bitterness of defeat. And Ben knows from experience which feels better. To the winner go the spoils, to the loser the ignominy of public humiliation. For a magical 30 seconds Ben feels nothing but elation as he is jumped on and jostled by the crowd of admirers surrounding him. In that moment he is oblivious to his rasping breath, racing pulse and profuse perspiration in the aftermath of his incredible effort. It is only when he looks up and sees his PE teacher approaching that he suddenly starts to feel sick.

Chapter 33

ADAM, FRESH FROM his nightmarish sexperience at the school, is sitting in the chair at the ophthalmologist's having the check-up he promised himself. He has already undergone a battery of tests on an earlier occasion to make sure he is suitable for laser blended vision treatment, and if he decides to have it done Professor Dan Reinstein at the London Vision Clinic is certainly the man to do it. His reputation precedes him. If it works, as it has done for people like Carol Vorderman, Bonnie Tyler and Phillip Schofield, it will mean an end to having to wear reading glasses or bifocals. Adam is forever losing them, breaking them or forgetting them. Neither are they getting any cheaper, and the best quality ones are costing him a small fortune.

Dan plans to skilfully shave a thin layer of cornea off the front of each eye, using a laser controlled by a computer. Adam's non-dominant eye will be adjusted for reading, and his dominant eye will be only minimally altered for excellent distance vision. The actual procedure, done under local anaesthetic, will only take a minute or two and leave Adam with two eyes each with slightly different visual fields. Quickly, however, his brain will learn to merge the two images and blend the vision so that he can see both near and distant objects clearly. And without glasses. Adam finds the prospect very exciting.

But why did he come to need glasses in the first place? What happens at the age of about 45 that makes us hold small print further and further from our eyes in order to read it?

As we get older, our eyes become less able to focus on near objects because the lenses within them become less elastic. Consequently, they cannot change shape sufficiently to become more curved so that they can focus incoming light rays on to the light-sensitive retina at the back of the eye. Our vision therefore becomes blurred. Laser blended vision, by changing the curvature of Adam's cornea, can overcome this and help the less elastic lens within to cope. At his age, his lens cannot become much less elastic so his vision will not change further and the procedure will work nicely. In a younger person, it wouldn't necessarily prove a permanent solution, because their lenses are likely to change in years to come.

If only Adam knew it, Ben's eyes were currently focused on the coveted Wilmslow Cup he had just been presented with for winning the mile race at the intercollegiate athletics meeting. Its solid silver surface is glinting in the bright spring sunshine and Ben has no difficulty at all reading his own name, which had been inscribed on it exactly one year before. He is enjoying his moment of glory, and is revelling in the congratulations from all around him. Even the arrogant Matt Roberts has eventually ambled over, put a hand on his shoulder and playfully called him a 'ratbag'. In the nicest possible way, of course. And with a grudging smile on his face. It is the nearest Ben will ever get to a full-blown compliment and he realizes the rivalry will continue for a good while yet.

Joel has smuggled some beer into the college and is drinking it now from a refillable sports drink bottle. If he gets caught drinking alcohol on college premises, he will be suspended. Paul, on the other hand, has vodka and Red Bull in his own bottle and Tom is puffing away on a cigarette. Ben is still coughing from the exertions of the race and can think of nothing worse.

Tom began smoking a couple of years ago and became one of the 157,000 11–15-year-olds who start the habit every year. Will he be one of the 80 per cent who become addicted? One of the 50 per cent of long-term smokers who eventually die from a smoking-related disease? Ben hopes not. He likes Tom. He hopes that it is just a part of his rebellious phase and a peer-pressure thing that will pass. What damage is he doing to his lungs every time he inhales? He's destroying the tiny hairs called cilia that project from the cells which line his airways and are designed to pulsate rhythmically upwards and take foreign particles like soot, viruses and bacteria with them. He is filling his alveoli with carcinogenic tar and producing fibrosis and scarring. He is damaging his DNA and making future tumours more likely to arise in his lungs and airways. He is producing a measurable constriction of his coronary arteries within just 30 seconds of inhaling and increasing his long-term risk of a heart attack or stroke. If he kicks the habit today, he'll soon run no more risk than someone who has never smoked. But if he carries on and becomes addicted, it could be curtains.

It's bad enough just breathing normal air, Ben is thinking, as he stifles another wheezy cough. He has lectured Tom on the subject before.

'Any epidemiologist will tell you, Tom,' he had told him after listening to a radio broadcast, 'you're more likely to die prematurely. Or suffer from erectile dysfunction at the age of about 20, which would be worse.'

'Nah,' Tom had said in response. 'I just don't trust these epidemiologists. They can make statistics say anything they want. In my view the definition of an epidemiologist is just a person broken down by age and sex.'

'Very funny,' said Ben. 'So doctors are all wrong, are they?'

'Yep,' Tom had agreed, 'and dangerous too. Think about this, Ben. There are 700,000 doctors in the United States and the accidental deaths attributed to them every year number 120,000. That's 0.171 accidental deaths per doctor. On the other hand, there are 80 million gun owners and the number of accidental gun deaths per year is 1,500. That's 0.0000188 accidental deaths per gun owner. Conclusion? Doctors are 9,000 times more dangerous than gun owners. Capito? Remember mate, guns don't kill people. Doctors do.'

Ben couldn't help laughing. But he still worries about Tom smoking and is not going to stop trying to get him to give up.

Chapter 34

WHEN EVE ARRIVED at her elderly parents' house a few minutes before Adam, it was a clear case of prosopagnosia. Joe didn't immediately recognize his own daughter's face. Coming from the classical Greek words *prosopan*, meaning face, and *agnosia*, meaning non-knowledge, prosopagnosia is a disorder where the ability to recognize faces is impaired while the ability to recognize other objects remains relatively intact. It is reported in people who have suffered head injuries and some seem to have a degree of it from birth. Since Stephen Fry, the English actor, comedian and writer, apparently has it, along with Scottish entrepreneur and philanthropist Duncan Bannatyne, Eve's father is at least in good company. But Joe's prosopagnosia is not due to trauma to the head or a childhood developmental aberration. Joe's condition is related to his gradual and disturbingly progressive symptoms of early dementia. It is slowly changing his behaviour and personality in a variety of ways and is the main reason why Adam and Eve have decided to pay a visit tonight.

Joe, squinting at Eve in the doorway, slowly took in the images of her hair colour and style, her clothes, body shape and voice. Eventually, he put it all together in his mind.

'Eve,' he said excitedly, 'come in, come in. You should have told me you were coming.'

Eve had told him about a dozen times that week. And three times just today. Dementia is a cruel and depersonalizing thing which robs people of their dearest relatives as it becomes more severe. Eve is saddened to see it becoming more obvious and is concerned about what the future holds for her father, who had formerly been such an intelligent and quick-witted man. Flashes of that brilliance still occasionally emerge, but they are becoming less frequent and are often blunted by inappropriate mood swings.

Sitting herself down in the drawing room, Eve is greeted by Grace, who has heard her talking to Joe in the hallway and has prepared a pot of tea and biscuits. Even she looks that bit older and more tired, frail and vulnerable than she did a week ago. Maybe it is the strain of coping with Joe and his increasing wandering and restlessness. For a while, they talk about the family and other everyday things and Grace enjoys an opportunity to gossip about the outside world and reminisce about the past a little too. Joe doesn't join in much, except to smile at moments when he feels he is expected to, or to repeat what Grace is already saying. What a terrible thing it is to grow old, Eve is thinking. It all seems to be so . . . downhill. She listens attentively as Grace explains about her stiff aching joints, her constipation and her fading vision. And about Joe's increasing deafness, unsteadiness on his feet and occasional urinary incontinence.

Most of all, Eve wants to hear more about Joe's deteriorating mental state, but with her father sitting there in the same room, that will have to wait for a more private moment. Instead they turn to the complex matter of their medicines. Grace is taking five different pharmaceutical drugs and Joe six. They all have to be taken at different times of the day, some with food and some not. To a certain degree, they interact with one another and cause a number of side effects, although it is difficult for the untrained Eve to be sure which symptoms are the result of the underlying medical conditions and which are due to the medicines. She makes a mental note to call their GP to see if she can arrange a meeting between the four of them so that the dosage and timing of her parents' medicines can be reviewed and better planned. This would perhaps enable Eve to invest in one of those gadgets she can buy at the pharmacy which store the tablets and spell out clearly exactly when each should be taken.

When does the process of ageing actually start? Eve is wondering to herself. When do we stop growing, maturing and developing and when do the trillions of cells which constitute our body start to deteriorate and decline? When are we in our biological prime and at what moment does that begin to wilt and involute? Has it already started for *her*, even though she is soon to bring new life into the world? What causes ageing? Can we live our lives in a way which will slow it down or make it less problematical, and could it ever be prevented altogether or even reversed?

There are many theories about the basis of ageing. Some people think it is just due to wear and tear. Your gums recede, you get 'long in the tooth', your teeth fall out and that's it. They don't grow back, they're not replaced and you look older. But it isn't as simple as that. Many tissues of the body are replaced and maintained all the time. Liver cells regenerate, as do skin cells and the cells lining our airways and our digestive system. The cartilage coating the ends of our bones can repair itself too, and fractures still heal however old we are. There is certainly a decline in our ability to repair and maintain our bodies as we grow older, but what fundamentally causes that? Is it the immune system acting like a biological clock as some scientists believe? Or is an ailing immune system a result of ageing and not the cause? Are we genetically programmed to age? If so, this would make some sort of sense as it would neatly explain the very different lifespans of different species and account for the fact that, in animal experiments at least, selective breeding of long-lived individuals can increase longevity by up to 50 per cent. If confirmed, could gene manipulation in humans at some point in the future help us to live until we are 120 or 150? Would we even want to? Then there is the 'error catastrophe' theory. This suggests that increased numbers of DNA mutations in cells lead to errors in the production of vital enzymes. This in turn allows abnormal substances to accumulate in the body and interfere with healthy cell function. One such abnormal substance is the age pigment lipofuscin. Joe and Grace both show it in the age spots all over the backs of their hands and forearms. As it happens, lipofuscin isn't harmful. But other abnormal substances produced this way in other cells of the body might be.

The likeliest cause of ageing is the finite reproductive capacity of dividing cells. It seems that individual cells are pre-programmed to divide and reproduce themselves about 50 times. After that, this capacity declines and the cells simply pack up and die. It is similar to what happens to your old washing machine or car. You can repair it for a while, and keep it going a while longer, but ultimately it will give up the ghost. This theory is borne out by what is observed in ageing cells and consistent with what happens in the rare inherited condition known as progenia. In this disorder, ageing is accelerated to the extent that a seven-year-old girl can easily look like a 90-year-old woman. This is because in progenia, the number of cellular divisions that take place before cell death is much smaller. In the ordinary process of ageing, chromosomal changes are identifiable too. Almost every cell in our body contains 23 pairs of chromosomes which carry the genetic code for our bodies. The tips of these chromosomes are called telomeres and scientists have discovered that, as we age, the telomeres become shorter. Their length is maintained by an enzyme called telomerase. If humans can find a way of boosting levels of telomerase, perhaps with a new medicine, maybe we will one day be able to live much longer.

Eve's daydream is interrupted by the doorbell. Joe, puzzled by what should really be a very familiar sound, looks plaintively at Grace for an explanation.

'It's Adam,' she explains. 'I'll let him in.'

Adam, in buoyant mood, strolls in and hugs Eve's parents warmly. After the briefest of catch-ups, he suggests that he and Joe take a stroll around the garden, leaving the ladies to chat.

Adam is fond of Joe, and the two of them have become very close over the years. So it's a shame to see the little garden, of which Joe was once so proud and in which he would previously slave away all weekend to make it perfect, falling to rack and ruin. This loss of interest and motivation is part of Joe's early dementia, Adam realizes, and observing him now, there doesn't seem to be any part of his body untouched by age. Within his brain, quite apart from the decline in his cognitive function, nerve cells have been lost, myelin sheaths, which insulate them and rapidly convey electrical impulses along the nerves, are absent. The neuro-transmitters involved in mood and the communication between nerve cells have diminished also, along with the synaptic receptors which respond to them. A total of 10 per cent of Joe's brain tissue has been lost since he was a young adult, and consequently his memory, his reflexes and his postural control and coordination have begun to desert him. His eyesight is bad enough, but his hearing has deteriorated markedly. Joe has presbycusis, the hardness of hearing of old age. Fifty per cent of over-70s suffer from it. Within his organ of hearing, the cochlea, the blood flow has diminished and many of the fine sensory hair cells which respond to different frequencies of sound, especially high-pitched ones, no longer function. So even if Adam raises his voice to make it louder, or shouts, Joe still struggles to grasp the meaning.

Joe is lucky that he still has Grace, but in many old people, this deafness simply adds to their loneliness, sense of isolation and depression.

At the back of each of Joe's eyes, the specific area of his retina responsible for detailed focusing, the macula, has partly degenerated. The region where millions of light-sensitive rods and cones are situated has become as threadbare as a worn-out old carpet. But the optician has said that Joe's type of age-related macular degeneration is dry AMD and caused by loss of actual cells rather than leakage of fluid and therefore can't be treated. Adam offers Joe a Werther's Original that he's extracted from his jacket pocket, but Joe declines. He doesn't get much pleasure out of food any more. It could be because at 86 he only has half the taste buds he possessed 50 years ago. As he reaches down to tidy some wayward netting over his once-prized strawberry patch, Adam notices how stiff and inflexible Joe has become, and how slow and deliberate his movements are generally. Is this all due to wear-and-tear arthritis – osteoarthritis – or is he also showing signs of Parkinson's disease or depression? Both conditions are common in old age and can lead to reduced, restricted and slower movements but for very different pathological reasons. Even the doctor has said it can be difficult to tell. Joe's old bones have actually done him proud, Adam reckons. He may be 3 inches shorter than he was 60 years ago due to the curvature of his spine and the narrowing of the cartilaginous discs between each of his vertebrae, but his bones have faithfully carried him around all these years and reshaped and remoulded themselves 100 times over. They have healed themselves when broken and produced hundreds of gallons of red and white blood cells from their bone marrow. In death, a skeleton is just an inanimate rigid framework of calcium phosphate. In life, it is a living, active, constantly metabolizing factory. Where would we humans be without bones? We would be as floppy as a beanbag. A puddle of skin and guts on the floor. Where would our heart and lungs be without the protective armour of our ribcage? How long would our brains last if they were not encased in our rigid skulls?

We have over 300 bones in our body at birth, and then 206 as an adult after some of them have fused together. Some of Joe's bones are still fusing together, but not in a good way. Arthritic changes and bony outgrowths in the small accessory joints of his lower back (the facet joints) have prevented all movement between some of his vertebrae, and bending down and straightening up again is a struggle. When he does finally straighten up, with Adam's help, he is breathless. He only has half the amount of elastic tissue in his lungs that he had when he married. That's the stuff which gives his lungs recoil and makes the actual process of breathing so much easier.

Poor old Joe, Adam is thinking. What else does he have to put up with? An enlarged prostate and a pair of kidneys that don't excrete the toxins in his blood as efficiently as they used to. His benign prostatic hypertrophy, common enough to cause urinary problems in one in three men over 50, is one thing, but, as it happens, buried deep within his prostate gland early cancer is lurking. Joe doesn't know this, but then neither does he need to. Prostate cancer becomes the most common cancer of all in men who live to the ripe old age of 90, and most men of this age will go to their graves with their prostate cancer undiagnosed, having died from some other cause. It's only in a few younger men that prostate cancer requires treatment, when the cancer is much more aggressive or associated with a familial predisposition. What else does Joe have to endure? Sluggish immunity which makes him more prone to infections like pneumonia. Increased auto-immunity which makes him more vulnerable to diseases like rheumatoid arthritis or blistering skin conditions such as pemphigoid.

As they trudge back towards the house and the warmth of the drawing room, Adam wonders what his own experience of ageing will be. He hopes that when the time comes, he won't have to suffer from all of the same depressingly chronic ailments that Joe is having to tolerate, and that he will just die peacefully in his sleep. Since like everyone else he spends about a third of his entire life asleep, if he's lucky, he has a one in three chance of that happening anyway. How quickly will these changes start to take place? Is he still in peak condition, or is he already, as he secretly fears, past his best?

A man is physiologically at his best in his middle to late 20s. So is a woman. At this time, all the cells in their bodies are functioning optimally. They can adapt and respond to changing circumstances better and quicker than at any other stage of their lives. Fertility and immunity are at their highest. The infections and cancers that young children and the elderly are prone to are less common. Their ability to respond to the stimulation of nerve impulses and hormones will never be better. But after the age of 30, these responses slowly begin to decline. Imperceptibly at first and sometimes postponed by a healthy lifestyle, inevitably and gradually it happens. Already, Ben and Poppy are teasing Adam about his 'man boobs', his male-pattern baldness and the greyness of his hair.

The fat deposition under his nipples is the result of reduced levels of exercise, eating too much generally, and over-indulging in alcohol. The changes in his liver which the alcohol is inducing are increasing the secretion of oestrogen which all men have to some extent, and making what little bit of breast tissue he has expand and grow. He's not quite ready for a sub-areolar mastectomy operation just yet, but he knows he should do something about it, such as losing some weight and cutting back on the booze. Otherwise, he'll be having to borrow one of Eve's bras.

His male-pattern baldness, inherited from his grandfather, obviously skipped a generation – his own father had a thick head of hair until he died in his late 70s – and neither Regaine foam nor Propecia tablets have seemed to make any difference. He has been contemplating a hair transplant for a while. It would involve taking a strip of skin from the back of his head and replanting the extracted hairs on to the bald area at the top and sides of his head, but he has thought better of it. Better to grow old gracefully, he thinks, especially with another baby on the way. The stress would almost certainly undo all the good work of the surgery, in spite of the tricologist's assurances that hairs transplanted from the back of the head do not respond to circulating testosterone in the same way and should therefore remain for ever. So Adam had declined the invitation to spend £7,000 on what was essentially a cosmetic procedure.

As for going grey, he is simply not producing the same amount of pigment in his hair follicles any more. Yes, he knows he could dye his hair and for a while he used to. But it was a bore, and so he stopped. Amazing, Adam thinks, that the average person starts with about 100,000 hairs on their head and the life span for a hair, before going into its cycle of shutdown and regrowth, is about three years. Since most people lose about 100 hairs every day, this means losing 35,000 hairs every year. In three years, that equals the total number of hairs on the head of an average male, but of course many of the hairs are growing back all the time. Except in Adam's case, he is thinking. It doesn't seem fair, but he accepts that there is nothing he can do about it.

One day soon, he reflects, he will end up like Joe. With a failing memory, unreliable judgement, limited concentration and sporadic confusion. He'll struggle to find the correct words for things and get into a muddle about the time of day. Already Joe has missed three appointments Eve has arranged for him at the doctor's. His anxiety, irritability and need to repeat the same old questions several times over are mounting. His rapid changes from happiness to crying are difficult to deal with. What is this dreadful disease known as dementia and why can't doctors manage it better?

Dementia at Joe's age is not uncommon. There are 700,000 people in the UK affected by it. But in an ageing population that figure is going to grow. It is estimated that it will rise to 1 million by 2025. It is rare in people under 65 but affects one in 20 over that age, and one in six over 80. Two-thirds of victims live in their own homes, and three-quarters of residents in care homes have dementia. Some of them, if they're physically robust, can live up to ten years with the condition. Adam shivers as he contemplates that depressing prospect.

Joe's doctor had explained about the changes which occur in the substance of the brain. An abnormal protein called amyloid is laid down in the outer layer of the brain in scattered conglomerations called plaques. Within them, nerve cells die and collapse in on themselves, which microscopically look like clumps or tangles. Macroscopically, magnetic resonance imaging (MRI) scans and computerized tomography (CT) scans show an overall shrinkage of the entire cerebral cortex.

To be fair to her, Joe's GP had been very helpful and supportive. She herself, she had said, had an elderly mother with dementia and so she totally understood and sympathized with Grace's concerns. She had carefully conducted an MMSE test (mini mental state examination), tactfully enquired about alcohol consumption and taken blood samples from Joe to rule out any treatable type of dementia. She had tested for anaemia, vitamin deficiency, inflammation, thyroid problems and kidney or liver function abnormalities. She had also requested brain scans that would help confirm the diagnosis and exclude low pressure hydrocephalus (water on the brain), which might also be a reversible cause of dementia, since the pressure which compresses and damages adjacent brain tissue can be surgically reduced by means of a drainage pipe called a shunt. Unfortunately, the end result of all these investigations was not favourable.

Joe is suffering from Alzheimer's disease, the most common form of dementia, and his symptoms are unlikely to get any better. He had been prescribed Aricept, a medication designed to increase levels of the neurotransmitter acetylcholine and, in so doing, improve among other things his memory. What a cruel end to such a wonderful life, thinks Adam. Joe had, and still does have at times of occasional lucidity, a most marvellous wit. Only last week he'd told Adam the story about the elderly couple in their 80s who had booked in to see a sex counsellor.

'How much to watch us make love in front of you so you can tell us what we're doing wrong?' the man says.

'Fifty pounds an hour,' replies the counsellor.

Whereupon the elderly couple made passionate love with no apparent problem at all. The counsellor duly told them so. Yet every week, week after week, they returned to carry out the same performance. After three months of this, the counsellor can take no more.

'Look. I've watched you make love for the last 12 weeks here in front of me and I can't find a single thing wrong with your lovemaking technique. Considering your age, I think you are amazing but I really can't justify taking any more of your money when I have nothing to offer in return.'

Joe had paused with excellent comedic timing and delivered the punchline with a wicked look in his eye.

'Well,' replies the elderly man, 'I'm married, so we can't go to my place. *She* is married, so we can't go to her place. The Travelodge hotel down the road charges £90 for a room, and the Holiday Inn on the other side of the street £120 for a room. You only charge £50 and we are reimbursed £40 of that from BUPA.'

Both Joe and Adam had fallen about laughing. And there was more.

'Let's go upstairs and make love,' says a 90-year-old woman to her husband.

'Pick one or the other,' replies the old boy. 'I can't manage both.'

Joe had been in his element.

'You know you're getting old,' he had told Adam, 'when your friends compliment you on your new alligator shoes, when you're actually barefoot at the time.'

'You know you're getting old when "getting a little action" means you don't have to take your laxative today.'

That previous week, Joe had been the witty, razor-sharp entertainer that Adam had always remembered. Today, he was a shambling wreck. Would the same degenerative neurological process pick on Adam? Would he be one of the one in six at Joe's age? Or would nature be kinder and let cancer claim him first?

After heart disease, cancer is the next biggest killer of the elderly. The ageing cells of the body, with all their DNA mutations and accumulated toxins, are much more likely to reproduce themselves unfaithfully and generate malignant changes, which in turn may spread to distant organs of the body through the myriad channels of the lymphatic system and blood. Some, like solid tumours in the brain, kidney or bowel, will simply spread out locally, eating away at surrounding healthy tissue and causing infection, obstruction or haemorrhage. Considering all the potentially carcinogenic influences elderly people have been exposed to throughout their lifetimes, perhaps it is not surprising. Chemical carcinogens in cigarettes and benzene. Physical carcinogens and irritants from dyes and asbestos. Other harmful causes of malignancy such as viruses, parasites and hormones. And how much radiation had Joe's parents been exposed to from X-rays and natural sunlight over the years? You only had to take a look at their skin to get an idea. There were so many wrinkles, creases, age spots and sores which wouldn't heal because they were, in fact, locally invasive, relatively benign skin cancers.

At least Grace has been lucky enough to avoid breast cancer. Affecting only one in 15,000 women aged 25, Grace's chances of developing it by the age of 85 were one in ten. Which would kill them first? Adam wondered. Dementia, heart disease or cancer? They aren't attractive choices. Maybe the side effects of the various medicines they are both taking will override all of them. Grace's outlook is far brighter than Joe's. She is fun, engaging and gregarious. She still has many friends who visit her or whom she visits. She still works on her own allotment, dances at the nearby social club, plays bridge for the local Women's Institute and retains a relentless curiosity for all things modern. Her TAVI has given her a new lease of life and the bilateral hip operations have been a runaway success. The pain, stiffness and immobility she had previously suffered from are now a thing of the past. For Grace, 85 is the new 65. Provided you are of sound mind. The safety and life-enhancing capacity of modern medicine means almost anything is possible.

Quietly sitting down again in the drawing room, the four of them chat for a while about the future. Suggestions of a live-in nurse or a move to a residential home together are rejected out of hand. But the occasional phone call from the doctor and the odd visit from the community nurse and meals on wheels would be just about acceptable. Joe and Grace could muddle through.

Eve realizes that one day in the not-too-distant future, Joe's and Grace's lives will inevitably come to an end. If fate is kind to them it might even happen peacefully in their sleep. Far better that, she thought, for all concerned, than the trauma and prolonged incapacity of a heart attack, stroke or cancer. The very thought of losing either of her beloved parents makes her shiver involuntarily, though. The notion of being without them is almost unbearable. Almost inconceivable. Their memory will live on within her for ever, she is sure, their souls eternally entwined with hers. But what actually happens at the moment of death? she wonders. Is the experience agonizing, frightening and ghastly? Or are there really beckoning angels and shafts of brilliant light gently guiding us towards heavenly peace and a new beginning?

In physiological terms at least, the dying process is one of accumulating cellular dysfunction. Whatever the cause, sudden or gradual, the failure of normal cell metabolism leads to a lack of energy and an inexorable need for sleep to conserve what little vitality remains. Appetite diminishes. Bowel movements decelerate. The kidneys produce less urine and the skin less sweat. The circulation becomes more feeble as blood pressure drops, leaving the skin cooler and a blue-grey colour with reddish-purple mottling where blood pooling occurs. The brain, semi-starved of oxygen and increasingly suffocated in the toxic by-products of sluggish biochemical processes, becomes confused, agitated and restless. The eyes, so sparkling and expressive in life, become dull, weary and sunken. Finally, the action of breathing itself becomes increasingly laboured and difficult. As sleep gives way to semi-consciousness, the rise and fall of the chest becomes shallower and less frequent, with accumulated secretions that cannot be swallowed and a build-up of fluid seeping into the air sacs of the lungs to produce that familiar bubbly sound known to clinicians as the 'death rattle'.

On one level, death can be regarded as the failure of several different organs all occurring simultaneously. Yet in most cases, it is a sequence, with each organ shutting down gradually, in turn, one by one. Clinical death takes place, and can be recorded for the purposes of a death certificate, when three things have been conclusively witnessed. There is no breathing. There is no heartbeat. The pupils are fixed and dilated. Within five minutes of the heart stopping, the cells of the brain, devoid of blood supply and the life-giving oxygen it contains, are no longer viable. Resuscitation at this point is no longer possible. Biological death has finally occurred.

Eve, staring straight ahead and transported to some dark, distant place, is suddenly jolted back to reality by a familiar voice.

'You shouldn't worry so much about old age,' Joe pipes up. 'It doesn't last long.' And Adam too, taken aback by this sudden flash of wit and unexpected lucidity, laughs out loud. He and Eve both realize they will have to settle on a compromise. It isn't the ideal solution, if it is anything remotely like a solution, Eve thinks. But it is a battle she is never going to win. She and Adam will support them, and be there for them as much as and whenever they can. And they can well understand Joe's and Grace's insistence on preserving their independence for as long as possible in their own home. The time will inevitably come when one of them has to live their life alone and a rethink will prove necessary. Until then, Adam and Eve promise each other that they will help Joe and Grace get as much as they possibly can out of each and every day that they have left together.

Chapter 35

RUBY HAD BEEN less than impressed to discover that her mummy and daddy wouldn't be home before she had to go to bed. The only good news was that she loved her sister Poppy to bits and she could twist her round her little finger. Poppy would let her sit up later than usual to watch TV. She could also clear up some of those questions to which her mummy didn't know the answer, such as why do you never see baby pigeons? Why are clouds only white and grey, and do Rufus and other dogs have a sense of humour?

Unfortunately, Ruby's evening hadn't gone entirely to plan. First she had an extended bout of hiccups. Next she started sneezing as soon as she cuddled up to Rufus and then, immediately after supper, she had been violently sick.

Hiccups can happen to anyone. Caused by an involuntary spasm in the diaphragm, that tent-like sheet of muscle separating the chest from the abdomen, it provokes a sharp intake of breath against a closed flap of tissue at the back of the throat called the glottis, and the characteristic 'hic' sound is made. Maybe it was because she had swallowed too much air with her fizzy drink that the hiccups had lasted for so long. Holding her breath hadn't stopped them, nor had pulling hard on her tongue or Ben coming up quietly behind her and scaring her half to death. It had taken Ruby the monumental effort of blocking off her ear canal with her fingertips while drinking half a pint of water through a straw in one go to finally stop the hiccups. By then she'd been hiccuping for half an hour. Not as long as Charles Osborne, who in 1922 had hiccupped for 68 years, but long enough all the same.

Twenty minutes later, Ruby had started sneezing. Some of the hairs from Rufus's wiry coat had got up her nose and irritated its lining. The chemical histamine had been released and messages had been sent along nerves to her brain which had triggered the reflex that had made her sneeze. Just before that happened, she inhaled deeply, her throat closed and the air pressure within her chest built up dramatically. As a result, the particles of fluid that flew out of her mouth and nose travelled a full 15 feet across the room at up to 110 miles per hour.

When Ruby had sneezed six more times in quick succession, Poppy acted. Pinching the middle of Ruby's upper lip just below her nose with her forefinger and thumb, Poppy applied pressure for about 30–40 seconds. Miraculously Ruby's sneezes had stopped and hadn't returned.

Finally, when neither Poppy nor Ruby was expecting it, another 30 minutes later, Ruby was violently sick just after supper. Poppy felt she had definitely earned her baby-sitting money. All Ben could do while Poppy cleared up the mess was play with his belly button fluff and moan about his ingrowing toenails.

Fantastic, thinks Poppy, hiccups, a sneezing fit and then puke. Finally, to cap it all, Ben's belly button fluff . . . that mixture of microscopic textile fibres, dead skin and bacteria. How gross.

'Why do boys have to be quite so disgusting?' Poppy had asked herself for the thousandth time. 'And when the hell is Mum getting home?' She was fed up and she wanted to go to bed.

Chapter 36

IN FACT POPPY'S parents had stayed longer than planned at Joe and Grace's house and had consoled themselves on the way home with a quick bite to eat and a fortifying drink at a favourite local restaurant. Eve had shed a few tears as she realized just how frail her father was becoming and then drifted off to sleep in the car as Adam drove home.

Unfortunately, she was soon woken up again with a completely dead left arm which she had to manipulate into a better position with her normal right arm. She had been propped up against the hard armrest on the passenger's door and this had pressed firmly against the main artery and nerve, cutting off the blood supply and interrupting the transmission of nerve impulses to the forearm and fingers. Her entire arm was cold and senseless.

It took two more minutes of intense pins and needles for normal sensation and function to be restored as the blood flowed back again. She had often experienced similar numbness in her legs and arms in bed when she had slept in an awkward position, but this was the first time it had happened in the car. Adam had just commented that if she had a little more flesh on her like he had, the extra padding would stop it happening. Eve, predictably, did not find the observation terribly constructive.

When they finally returned home they found Ben and Poppy deep in a TV movie and Ruby curled up asleep next to Rufus on the floor. But it was 11 o'clock at night and definitely time to hit the sack.

Lying in bed that night, next to a slumbering and contented Eve, Adam is methodically going through the events of the day in his mind. Eve's skin is warm and silky, perfectly smooth, soft and completely free of blemishes. Perfect at any time, her pregnancy has seemed only to improve it further. She must have been exhausted, he thinks, because she fell asleep within moments of her head touching the pillow. There had been a hint of soft snoring for a second or two as she initially lay face up, allowing her tongue to fall towards the back of her throat and partially obstruct her airway. Then she had suddenly turned on her side, enabling her tongue to fall forward again, and her breathing had become silent. Only the gentle rise and fall of her beautifully sculpted shoulders told Adam that she was breathing at all.

All in all it hadn't been such a bad day. Poppy had been amazing, collecting Ruby from her primary school, bringing her home and feeding her while her parents were out visiting Joe and Grace. Then she had had to deal with that triple whammy of Ruby hiccuping, sneezing and then throwing up. A better, more wonderful daughter Adam couldn't have hoped for. He knew that her biological mother, Rawaa, still consistently in Eve's thoughts, lived on in her spirit like a shining, inextinguishable beacon.

Ben, who had retained the Wilmslow Cup, looked exhausted as he sprawled across the sofa and Adam and Eve, who have always encouraged their children to participate in competitive sport, couldn't have been more proud of him.

Eve's pregnancy is progressing without complications, and the newest little Enniman, the fledgling Enniman, at 12 weeks' gestation is well on the way to becoming a viable human being. In another four weeks it will be time to tell their friends and let the world know about the baby's existence.

Adam's talk today had been, well, interesting, but he won't be doing it again any time soon, he reflects as he lies awake.

Everyone else is now fast asleep and recharging their batteries for the next day. Adam himself is very tired and weary too. The trouble is, he just doesn't seem to be able to drift off to sleep. It has been an increasing problem for him over the last few weeks and his resultant daytime drowsiness has not been easy to cope with.

He has tried power napping a number of times. He knows the principle perfectly now. First, drink a strong coffee. Next, set the alarm on his mobile phone for 18 minutes precisely and no more. This is so that he dozes but never gets into deep REM sleep, which would leave him feeling groggy, disorientated and irritable upon waking. He then kips for the allotted 18 minutes and just as the alarm wakes him up, the caffeine is kicking in, making him feel more alert and leaving him refreshed. Bingo. The beauty of the power nap. Unfortunately for Adam, it doesn't seem to work. It still makes him feel grumpy and irritable when he wakes up, so he has abandoned it.

Adam, like 30 per cent of the population, including Eve's parents, suffers from chronic insomnia. Why can't he sleep as soundly as the rest of the family? he wonders. Why can't he enjoy the physical restoration and mental repair that sleep provides? Sleep is so important. It is a time when the brain can assimilate fresh information, when the emotional content of it is assessed and processed and when memory is vitally updated. Without it we suffer sleep deprivation and the physical and psychological distress that this causes. Sleep deprivation, after all, has often been used as a form of torture. But the more he worries about drifting off to sleep, the harder it seems to come to him. And that, in a nutshell, is Adam's problem. If he were to ask anyone who sleeps wonderfully what they do to achieve it, they would invariably tell him – *nothing*. They just let it happen. The more Adam frets about not sleeping, the worse he makes it. All those tricks like counting sheep, reading a book, having a hot drink or imagining a stroll on a sunny beach count for nothing. Nor would Adam's doctor entertain the idea of prescribing sleeping pills.

'If I prescribed you sedatives, they'd certainly help for a few nights,' he had said, 'then your brain would become tolerant to them and you'd need a bigger dose to achieve the same effects.'

He was right. The most commonly used benzodiazepine

sedatives, just like general anaesthetics, depress the reticular activating system in the brain stem and promote drowsiness and sleep. But the brain is clever. It quickly realizes that there are external influences at work and it does everything in its power to fight them. The result is that, as soon as the sedatives are discontinued, the brain becomes hyper-alert and even more active than before.

Adam's doctor was clearly not going to play ball. He'd have to find some natural solutions of his own. Lately, he had tried eating earlier before going to bed. He'd hoped that the process of digestion would not then boost his metabolism and increase his body temperature while he lay in bed. He'd made the bedroom quieter and darker and had studiously avoided caffeinated drinks after 7pm. He'd tried to incorporate more exercise into his working day. He'd attempted to get into a more regular sleeping routine. He'd read books in bed, he'd had the TV on quietly and he'd started playing relaxing music on his MP3 player and had invested in a new, state-of-the-art, body-contouring Tempur mattress. He's even made efforts to cut back on alcohol – as the doc had told him that it interferes with his normal sleeping patterns. He acknowledges that he'd suffered from a hangover for half the day, but surely it couldn't still be doing that twenty-four hours after his drinking spree? He turned over irritably to look at the clock: 1.15am. Tonight, as it so often has been lately, sleep is proving an elusive and capricious companion.

With the onset of sleep now slowly beginning to wash over Adam, he is going into his first cycle of NREM sleep as he becomes increasingly drowsy. At this point, his brain waves are becoming deeper and slower. His brain activity and his body's metabolic rate are at their lowest ebb of the day. Dreams, if he has them at all, are infrequent. He might experience the occasional myoclonic jerk – as Eve did earlier when her arms and legs kicked out as if she was still conscious and reacting to imaginary events – but as his brain waves slow further, this will stop.

As Adam slips deeper into sleep this first cycle of NREM sleep is suddenly interrupted. His brain waves become hyperactive and frenetic. They resemble the brain waves he would have if he was wide awake. The difference here is that he is now deeply unconscious and the muscles of his body are totally paralysed. Adam is now in REM sleep when his temperature rises as the blood flow in his brain is increased. Bursts of nerve cell activity in his brain stem shoot towards his visual cortex in the occipital lobe of his brain. His eyes are moving rapidly from side to side and he is dreaming.

Adults like Adam will spend about 20 per cent of their total sleep time in REM sleep. For babies, it is nearer to 50 per cent. For him this first episode of REM sleep has begun at 2.45am and lasts about ten minutes, but Adam will experience four or five more episodes of REM sleep throughout the night, each lasting a little longer than its predecessor. The last one, just before he wakes again at 7am the next day, will last one hour and he may or may not remember what he has dreamed about.

Some scientists believe we sleep and dream to cement experiences into long-term memory; in other words, we dream about things that are worth remembering. Others are convinced we dream about things worth forgetting – to eradicate overlapping memories that would otherwise overwhelm our brains. The truth is, nobody actually knows.

To a large extent, the exact purpose of sleep remains an enduring mystery which eludes the most academic of sleep specialists. Neurophysiologists attempt to describe it as decreased activity in the reticular formation of the brain, brought about by the reduction in excitatory neurotransmitters like noradrenaline, which in turn inhibits activity in the thalamus, hypothalamus and cerebral cortex. But even that is a simplification.

William Shakespeare penned a different and more poetic description some four hundred years ago. He wrote:

> The innocent sleep,
> Sleep that knits up the ravell'd sleeve of care,
> The death of each day's life, sore labour's bath,
> Balm of hurt minds, great nature's second course,
> Chief nourisher in life's feast.

The death of each day's life is perhaps as near to a perfect description of sleep as anyone might ever achieve. At this moment all five members of the Enniman family are soundly asleep and oblivious to the outside world. For Adam and the rest of his family, the birth of the next day's life is just around the corner.

Acknowledgements

WATSON & CRICK ALLEGEDLY unravelled the secrets of DNA over regular drinks at the pub. In similar fashion, Doug Young, my editor, David Miller, my literary agent, and I came up with the idea of unravelling the secrets of the entire human body and everything it does within a 24-hour timeframe over a boozy lunch.

So first, I need to thank them. Then I need to thank all my friends and colleagues who responded to my emails asking them 'what do you want to know about your body and what it achieves in a day?' Chapter 30 is consequently mainly dedicated to them, focusing as it does on questions about sex. But they also asked thousands of other questions unrelated to this popular obsession – amongst them, why do we yawn, why does urine change colour, what is the best way to treat a hangover and what happens at the very moment we die?

My very good friend Steve Wright and his side-kick Tim Smith started the ball rolling with a myriad of random medical questions over the years whenever I was a guest on his massively popular Radio 2 show, but Christina McDowall, Charlie Leslie, Michelle Porter, Ali Lutz, Helen McMurray, Trevor Lejeune, Henry Vines and Annemarie Leahy have all proved particularly inquisitive.

I'd like to thank Carol Clements for her patience with typing the manuscript, my PA Kim for planning my diary and keeping me to the deadline, and Dee and all of my family for putting up with me when I was immersed in the book and couldn't come out to play.

Finally, a thank you to all of my patients over the years, whose bodies have behaved in peculiar ways, and who have subsequently asked me 'why?'.

Index

More from Dr Hilary Jones . . .

What's Up Doc?

For Dr Hilary Jones, the question 'What's up doc?' has been asked of him ever since he qualified as a doctor at the Royal Free hospital in London over thirty years ago.

As a junior medic patients used to ask him 'What's up?' when he prodded their bellies for signs of appendicitis. On the GMTV sofa presenters ask him 'What's up?' with the latest actress who has developed the typical tell-tale signs of anorexia nervosa. In the tabloid newspapers he's asked to comment on what's up with the premier league footballer who purports to suffer from sex addiction. On the radio he's asked 'What's up?' with the health of society in general, suffering as it does from epidemics of obesity and binge drinking.

On a more everyday basis, in the GP surgery people ask him about unexplained lumps in their neck, or whether a pigmented mole is suspicious. Colleagues at work stop him in the corridor and say 'Can I just ask you about my child's leukaemia' or 'My mum's dementia?' At dinner parties people ask him about their haemorrhoids, or in pubs on the various merits of vasectomy. He's even been approached by complete strangers in dimly lit streets eager to hear his take on methadone and whether or not the NHS should freely supply it.

And they ask him what Lorraine Kelly is really like, of course . . .